Exploring The BUILDING BLOCKS of Science

Book 1

STUDENT TEXTBOOK

REBECCA W. KELLER, PhD

Illustrations: Janet Moneymaker

Copyright © 2014 Gravitas Publications, Inc.

All rights reserved. No part of this publication may be reproduced, stored in a retrieval system, or transmitted, in any form or by any means, electronic, mechanical, photocopying, recording, or otherwise, without prior written permission from the publisher. No part of this book may be used or reproduced in any manner whatsoever without written permission.

Exploring the Building Blocks of Science Book 1 Student Textbook (softcover)
ISBN 978-1-936114-30-6

Published by Gravitas Publications Inc.
Real Science-4-Kids®
www.gravitaspublications.com

Contents

Introduction

Chemistry

Biology

Physics

Astronomy

Conclusion

Chapter 1 What Is Science?

Introduction

1.1 Introduction

Have you noticed that ice always floats in a cup of cold water? Do you think about how a flower blooms or why leaves are green? Do you wonder what happens to the Sun at night? Have you thought about how wings help a bird fly?

All of these questions are about how things work. Science provides us with a method for studying how things work.

Science has several different parts. The first part of science is making observations and asking questions.

The second part of science is thinking of ways to answer these questions. To answer questions, scientists do experiments to test ideas and collect information. Testing ideas makes science different from all other types of study.

The last part of science is comparing ideas with other scientists and sometimes even arguing about the answers to the questions!

In summary, scientists observe and ask questions, do experiments, and compare their experiments to those of other scientists. This is what science is all about!

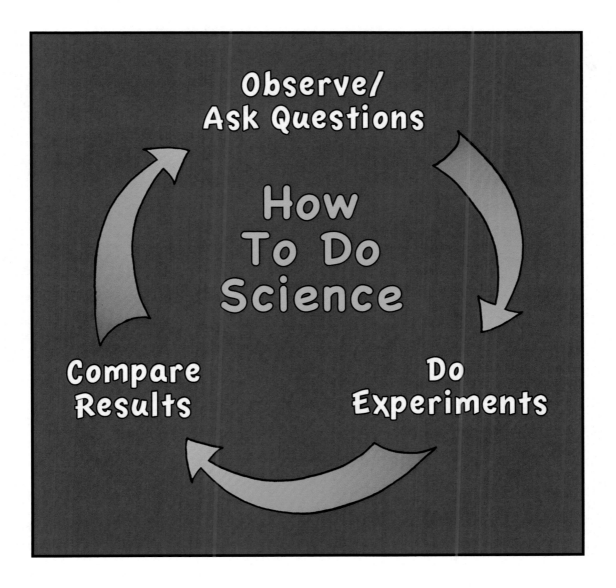

1.2 History of Science

A very long time ago students did not study "science." They might have studied Latin, music, or poetry, but there were no books on science and there were no scientists. So where did science come from?

The history of science is long. It's not easy to say exactly where and when science started.

We know that ancient people learned how to grow food, studied the stars, knew about planets, and even understood enough chemistry to refine gold and silver so they could make things. Even though ancient people didn't formally study science, they contributed ideas and knowledge that helped lead to what we now call science.

The science that we study now is like a story that has been written and rewritten many times. Many people over many thousands of years have contributed to the story of science.

Today, the way we explain how things work is very different from the way ancient people explained how things work. Over time scientists have changed the story.

For example, people used to believe that the Sun moved around the Earth. But today we know that Earth moves around the Sun. By helping us understand gravity, motion, and how planets move, scientific discoveries have changed the story of how the Earth moves.

It is important to keep in mind that stories for how things work continue to change. Even today we don't know everything about the world around us or everything about how things work.

As scientists learn more and more, they provide better and better explanations of how things work.

These explanations may either change or confirm ideas that we believe are true. Or they may lead to the development of surprising new ideas. This is how new scientific discoveries are made and how science helps us understand the world around us.

1.3 Building Blocks of Science

Modern science is a combination of different scientific subjects. A scientific subject is a specific area of science that someone studies.

Although there are many different scientific subjects, five subjects make up the foundation of all science:

chemistry, biology, physics, geology, and astronomy.

It is easy to think about studying science if we think about the different scientific subjects as building blocks that fit together and rely on each other.

To understand science a student needs to study at least some of the important topics in each of these five scientific building block subjects. For example, atoms are studied in the chemistry building block, force in the physics building block, and plants in the biology building block. These are three different essential topics in three different building block subjects.

Often, a scientific topic in a particular scientific building block cannot be understood without knowing about topics in other building blocks. For example, in astronomy to  understand how planets move around the Sun, a student must learn about gravity. Gravity is a topic typically covered in physics. So physics is a building block necessary for understanding astronomy.

Another example is understanding how plants make food. Both chemistry and physics are required for an understanding of the biology of how plants make food. Chemistry and physics are both building blocks that are essential for understanding the building block of biology.

In this book you will be introduced to many different science topics. You will learn about atoms in chemistry, cells in biology, energy in physics, the Earth in geology, and Earth's orbit in astronomy. You will see how the five scientific building block subjects fit together to provide an understanding of different science topics.

When the five essential building blocks of science are put together, a solid foundation for understanding science is formed.

1.4 The Scientific Method

To do science, a scientist follows a certain set of steps. These steps are called the scientific method.

The first and most important step is to make good observations. A scientist doesn't just think about things but rather looks at things in detail. It might be important to look at how tall something is or what color it is. How heavy something is or whether it is hot or cold might be things that need to be measured. During all of these activities, the scientist is making observations. Making good observations is the first and most essential step in the scientific method. A scientist can't study something without having carefully observed it first.

The second step in the scientific method is asking a question about what has been observed and then turning that question into a hypothesis. A hypothesis is a statement about something. For example, a question might be "How heavy is an elephant?" Turning that question into a hypothesis, it becomes, "An elephant is heavier than a kangaroo."

The third step in the scientific method is creating an experiment to prove or disprove the hypothesis. Is an elephant heavier than a kangaroo? The scientist doesn't

make any assumptions about the answer but performs an experiment to find out. A big scale might be used to measure how much each animal weighs. The scientist might weigh the kangaroo first and then the elephant, or the elephant might be weighed first and then the kangaroo. The scientist designs the experiment to prove or disprove the hypothesis.

By measuring how much an elephant weighs and how much a kangaroo weighs, the scientist can then move to the next step of the scientific method, collecting results.

What was the outcome of the experiment? What were the actual numbers observed on the scale?

The results of the experiment that was performed are written down. Writing down the numbers means that the scientist doesn't have to guess about the results of the experiment. It is important that the information is recorded exactly as it is observed and that no guesses are made about what the answer might be. By not guessing what the answer might be, a scientist remains open to new discoveries.

Finally, the last step of the scientific method is to draw a conclusion. What did the information show? Was the elephant heavier than the kangaroo or was the kangaroo heavier than the elephant?

The conclusion comes from the results of the experiment. Based on the results, the scientist draws a conclusion and shows that the hypothesis has been either proven or disproven.

1.5 Summary

- Science is a way to study how things work.

- Scientists make observations, ask questions, do experiments, and compare their experiments with other scientists.

- There are five scientific subjects that make up the essentials of all science: chemistry, biology, physics, geology, and astronomy.

- To do science, a scientist follows a certain set of steps. These steps are called the scientific method.

Chapter 2   What Is Chemistry?

Chemistry

2.1 Introduction

In Chapter 1 you learned that science is a way to study how things work. You also learned that science is made up of five different scientific subjects: chemistry, biology, physics, astronomy, and geology. These five subjects are the building blocks for all science.

In this section we will explore the building block called chemistry. Chemistry investigates what physical things are made of and the ways in which they change. Everything you can see, smell, or touch is made of "something." Scientists call this "something" matter. Chemistry is the study of matter.

2.2 History of Chemistry

A long time ago, students didn't study chemistry in school. Today, chemistry is an important part of all science, and students all over the globe study chemistry.

But where did we get chemistry?

Without knowing about chemistry, ancient people still did lots of chemistry. Ancient people used chemical processes every day.

A chemical process is any activity that involves some kind of chemistry.

For example, when ancient people treated animal furs so that they could be used as clothing, they were doing chemistry.

When ancient people created paints to draw pictures on walls, they were doing chemistry. And when ancient people made bowls or coins from metal, they were doing chemistry.

Ancient people did a lot of experimenting to learn about the world around them. Experimenting just means testing ideas and observing what happens.

For example, by experimenting, ancient people learned whether or not something would burn or melt. They learned if something tasted sweet or bitter. They learned that some things change color when added to other things. They even learned that some plants could be used as medicine. All of this is chemistry.

However, people didn't start thinking about chemistry as a science until the late 16th century. Around this same time, other scientific subjects, such as physics and astronomy, were also being developed. Also at this time, there were lots of people who did not do any experimenting at all, but just thought about things.

Thinking about how things work and doing experiments are two different ways to learn about the world.

In the late 16th century, thinkers began thinking about experimenting. Eventually thinking and doing were blended together, and these two ways of studying the world turned into what we call chemistry.

2.3 Modern Chemistry

Thinking about how the world works and doing experiments to observe how the world works are both important to modern chemistry.

We will see in Chapter 3 that matter is made of atoms. But atoms are so small that they can't be seen with our eyes. So if we can't see them with our eyes, how do we know they really exist?

Modern chemists do experiments to show that atoms exist. From these experiments they can learn a great deal about atoms. For example, by doing an experiment a chemist might find out how large an atom is, how fast it can move, or how heavy it is.

Modern chemists also think about atoms and make models to explain how atoms behave. A model is a good guess about how something works. Models can be created as writing, drawings, or computer programs, or they can be built from different materials. A model may not be completely true, but good models help chemists understand how things might work.

In Chapter 3 you will see a model for atoms. Each atom is drawn with a face and arms. Real atoms don't have arms, but that's OK. The models of atoms used in Chapter 3 will help you understand how atoms work.

2.4 Everyday Chemistry

You use chemistry every day, but you probably don't know it.

When you brush your teeth in the morning, chemicals in the toothpaste help clean your mouth. This is chemistry.

If you have a tummy ache, your mom might give you some medicine to make you feel better. This is chemistry.

If you want ice cream, your big sister might put gas in the car to drive you to the ice cream store. This is chemistry.

Even when you use watercolors to paint a picture, you are using chemistry.

Chemistry is everywhere! In the next few chapters you will learn more about chemistry and how we use chemistry every day.

2.5 Summary

○ Chemistry is the study of matter.

○ Chemistry is both thinking about how things work (making models) and doing experiments.

○ Chemistry happens every day, and we use chemistry in many different ways.

Chapter 3 Atoms

3.1 Atoms

Have you ever wondered if the Moon is really made of green cheese?

Have you ever thought the clouds might be made of cotton candy?

Have you ever wanted to know
what makes carrots
orange...

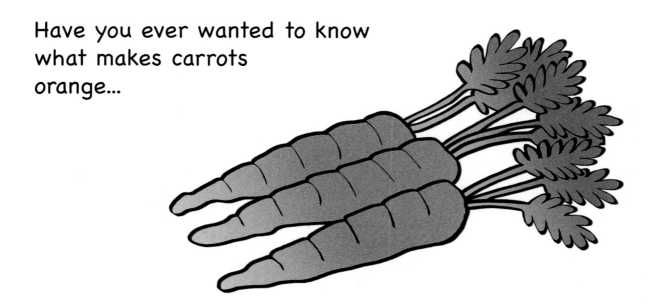

or peas green?

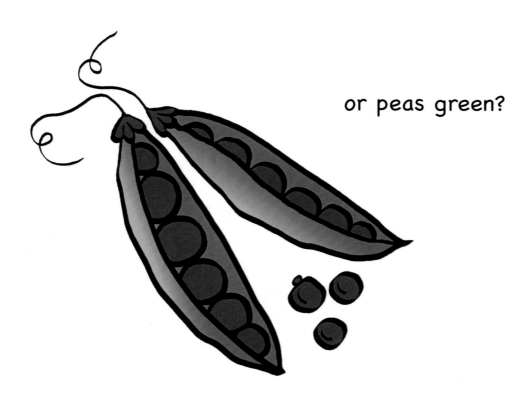

Have you ever wondered why brussels sprouts couldn't taste more like sweet cherries, or asparagus taste more like candy canes?

Everything around us has a different shape or flavor or color because everything around us is put together with different combinations of atoms. An atom is the smallest unit that makes up matter. Atoms can combine with one another in different groupings to make different substances.

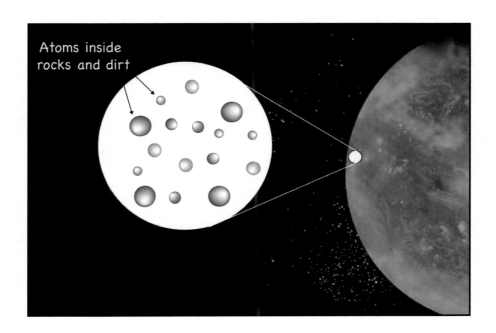

Atoms inside rocks and dirt

The Moon is not really made of green cheese. It is made of the kinds of atoms that are found in rocks and dirt.

Clouds are not made of cotton candy, but of the kinds of atoms that are found in air and water.

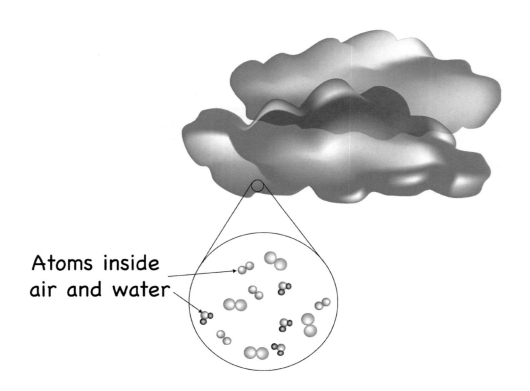

Atoms inside air and water

Carrots are orange because their atoms are arranged in a way that makes them orange.

Peas are green because their atoms are arranged in a way that makes them green.

Brussels sprouts and asparagus don't taste sweet like cherries or candy canes because the atoms inside brussels sprouts and asparagus are not arranged in a way that makes them sweet.

3.2 Types of Atoms

There are over 100 different atoms. Carbon, oxygen, and nitrogen are the names of a few of these atoms.

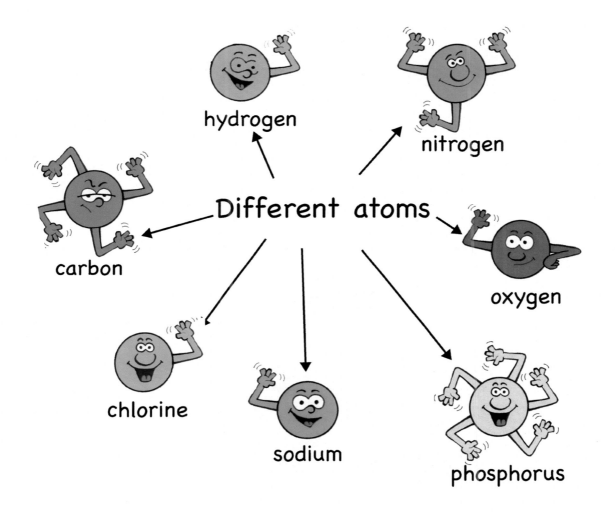

Atoms are very tiny. They are so small that you can't see them with your eyes.

Even though we can't see atoms with our eyes, we can make models of them. Remember from Chapter 2 that a model may not be exactly true but it can help chemists understand how things work. We can make models of atoms by drawing them as dots or little balls or by giving atoms colors and shapes.

In this book, atoms are modeled as balls with "arms" to help show how atoms hook to other atoms. Even though atoms don't really have arms, it's a great way to think about how atoms hook together.

Simple Model of a Silicon Atom

3.3 Atoms Are Similar

All atoms are made of protons, neutrons, and electrons.

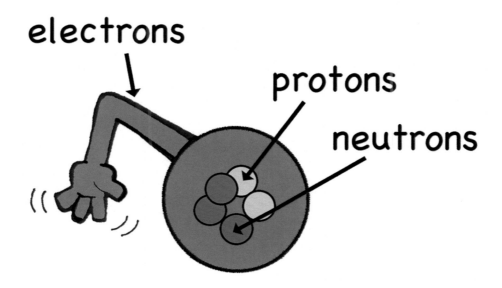

Protons, neutrons, and electrons are the basic parts of atoms. The protons and neutrons are in the center of an atom, and the electrons are on the outside.

In our model, the protons and neutrons are shown inside the ball of the atom, and the electrons are shown on the outside as the arms.

The arms in our models represent only the electrons that can help hook an atom to another atom. Some atoms have additional electrons that won't hook to another atom, and these electrons are not shown in our models.

3.4 Atoms Are Different

Atoms are different from each other because they have different numbers of protons, neutrons, and electrons.

For example, hydrogen only has one proton and one electron. It doesn't have any neutrons.

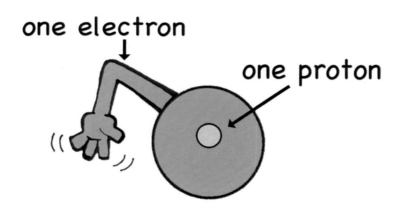

Model of a Hydrogen Atom

Carbon has six protons, six neutrons, and six electrons. However, only four of those electrons can help carbon hook to other atoms, so in our model carbon has only four arms.

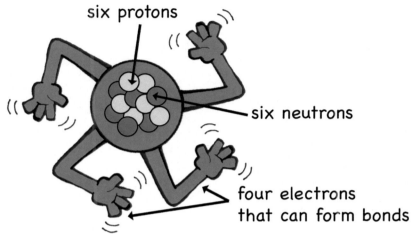

Model of a Carbon Atom

Uranium has 92 protons, 92 electrons, and 146 neutrons! The electrons in uranium can hook to two atoms, four atoms, or even six atoms!

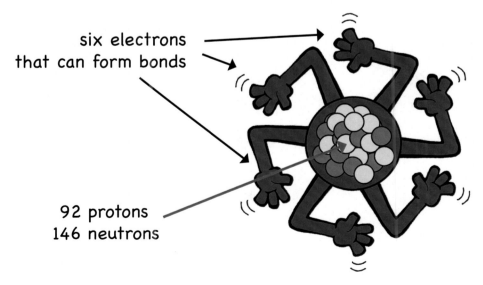

Model of a Uranium Atom

Because atoms are made of the same basic parts, all atoms are similar to each other. But because atoms are also made of different numbers of those parts, atoms are also different from each other.

Everything you can touch with your fingers, see with your eyes or smell with your nose is made of atoms. And all of these things are different from each other because their atoms are arranged in different ways.

3.5 Summary

○ Everything we touch, taste, see, or smell is made of atoms.

○ Atoms are very tiny things we can't see with our eyes. Atoms are the smallest units of matter.

○ All atoms are made of protons, neutrons, and electrons.

○ Atoms are different from each other because they have different numbers of protons, neutrons, and electrons.

Chapter 4 Molecules

Chemistry

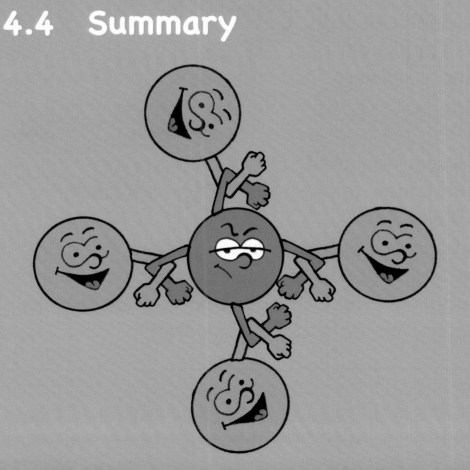

4.1 Introduction

In Chapter 3 we saw that everything is made of atoms. The Moon is made of atoms. The clouds are made of atoms. Carrots, peas, brussels sprouts, and asparagus are all made of atoms.

We also saw that all atoms are made of protons, neutrons, and electrons. We saw that atoms are different from each other when they have different numbers of protons, neutrons, and electrons.

But how do atoms make so many different things?

If we think of atoms as building blocks, we can begin to understand how atoms can make so many different things.

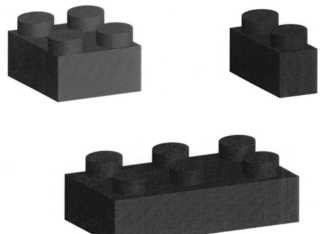

Building blocks are designed to hook to each other to make toy buildings, toy

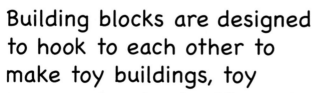

cars, or toy boats. If you look carefully at a building block, you can see that the pegs of one block fit into the holes on another block. By hooking building blocks together, different objects can be created.

In the same way that building blocks hook together to make toy cars or buildings, atoms hook to other atoms to make clouds, carrots, peas, candy canes, and the Moon!

4.2 Atoms Form Molecules

When one atom hooks to one or more other atoms, they form a molecule.

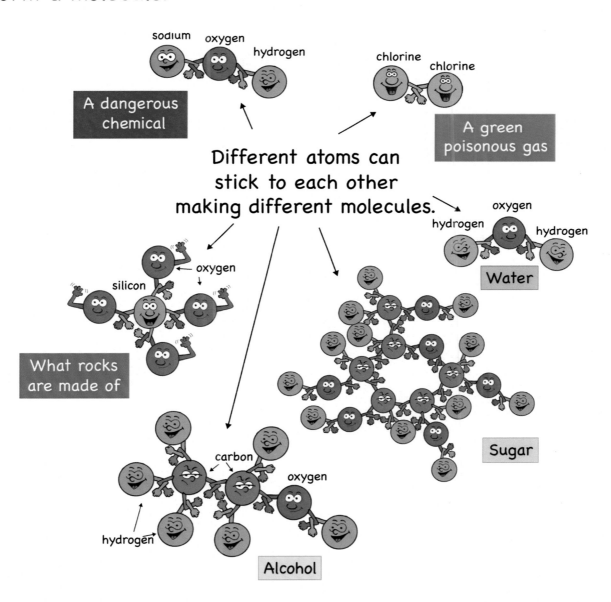

sodium   oxygen   hydrogen

A dangerous chemical

chlorine   chlorine

A green poisonous gas

Different atoms can stick to each other making different molecules.

oxygen
hydrogen   hydrogen

Water

silicon   oxygen

What rocks are made of

carbon   oxygen

hydrogen

Alcohol

Sugar

Sometimes only a couple of atoms hook together to make a molecule.

For example, table salt is made of just two atoms—sodium and chlorine.

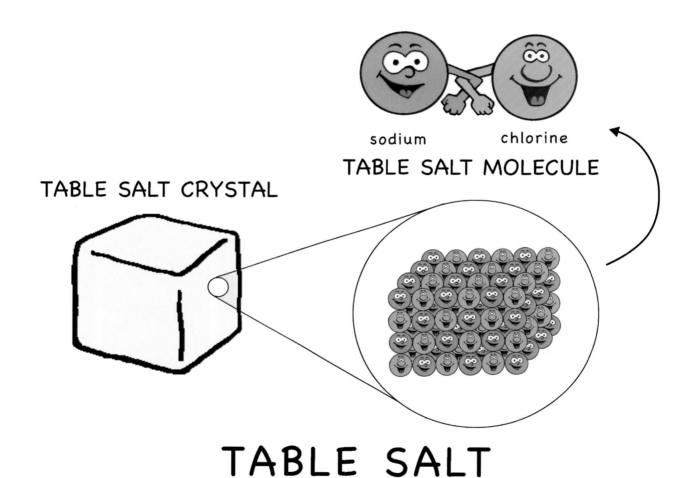

sodium   chlorine
TABLE SALT MOLECULE

TABLE SALT CRYSTAL

TABLE SALT

Water is made of three atoms—two hydrogen atoms and one oxygen atom.

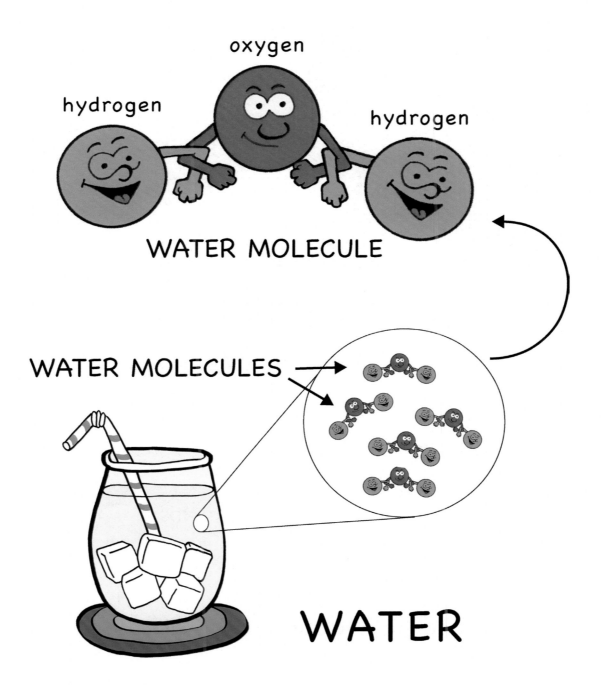

oxygen

hydrogen

hydrogen

WATER MOLECULE

WATER MOLECULES →

WATER

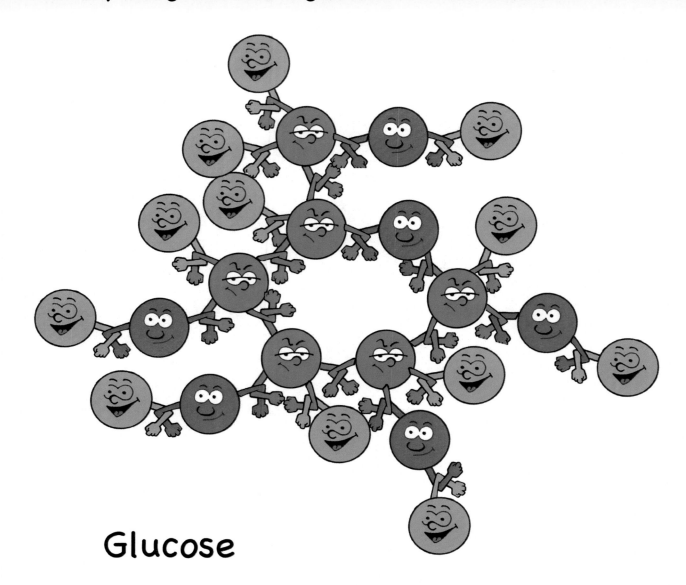

Glucose

Other molecules are made of more atoms.

Glucose, a type of sugar, is made of 24 atoms. It has 6 carbon atoms, 6 oxygen atoms, and 12 hydrogen atoms. (Count them in the above illustration.) Glucose is a type of carbohydrate. Carbohydrates are made of carbon and water. Carbohydrates give your body energy. (See Chapter 12).

Some of the molecules in your body, such as proteins, are made of thousands and thousands of atoms.

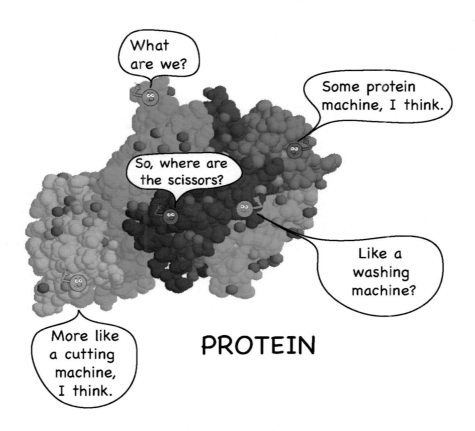

PROTEIN

4.3 Atoms Follow Rules

Atoms hook to other atoms by following rules!

For example, each hydrogen atom can only hook to one other atom. Hydrogen cannot hook to two atoms, three atoms, or more than three atoms.

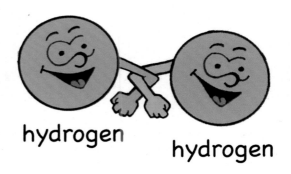

Oxygen cannot hook to more than two atoms. Nitrogen can hook to one, two, or three atoms, but nitrogen cannot hook to more than three atoms.

oxygen

nitrogen

Carbon can hook to one, two, three, or four atoms, but carbon cannot hook to more than four atoms.

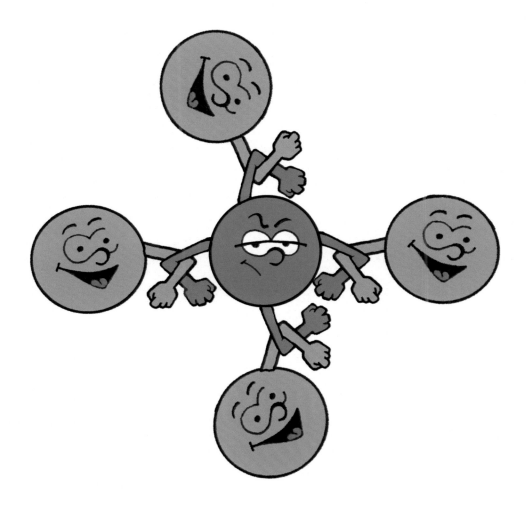

Many different shapes and sizes of molecules can be made with atoms. However, atoms always obey rules when they make molecules. Following the rules means that table salt will always be table salt and sugar will always be sugar!

4.4 Summary

○ Atoms hook to other atoms to make molecules.

○ Two atoms or many atoms can hook together to form a molecule.

○ Atoms have to obey rules when making molecules.

Chapter 5 Molecules Meet

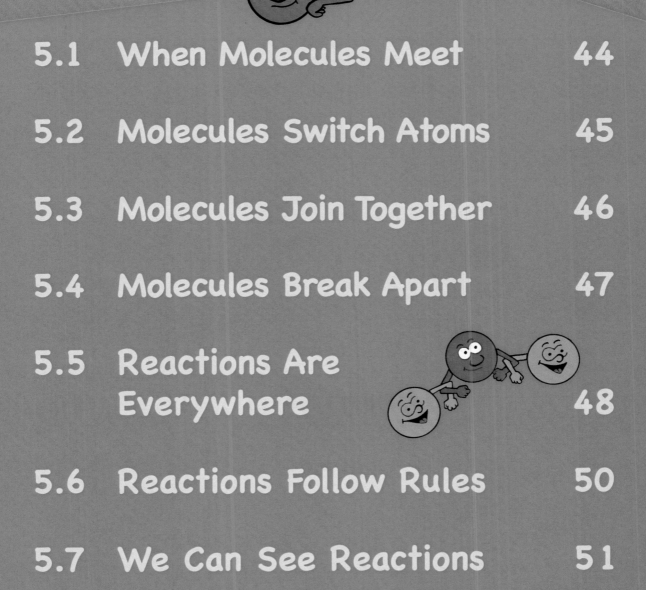

Chemistry

5.1 When Molecules Meet

In the last chapter we saw that atoms hook together to make molecules. We also found that atoms must obey rules. Each atom hooks to other atoms in its own way.

But what happens when one molecule meets another molecule? What do they do? Do they change, or do they stay the same?

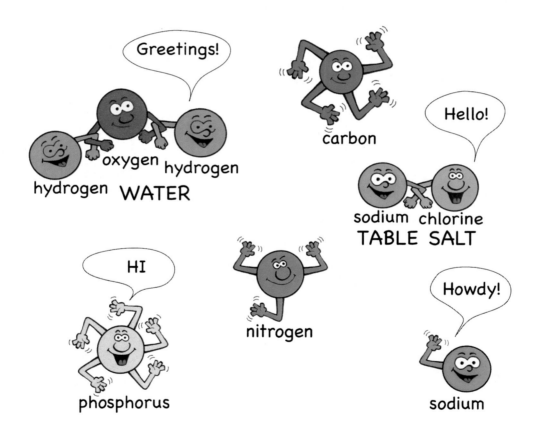

Sometimes when one molecule meets another molecule, they react. This means that something changes in the way the atoms are hooked together.

5.2 Molecules Switch Atoms

Sometimes molecules react by switching atoms (exchanging partners). In the example below, two molecules meet and trade atoms. As a result, two new molecules are made.

1 Two molecules meet.

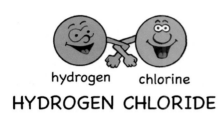

HYDROGEN CHLORIDE

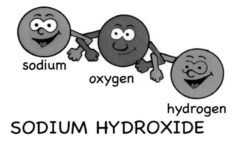

SODIUM HYDROXIDE

2 The hydrogen atom and the sodium atom switch places.

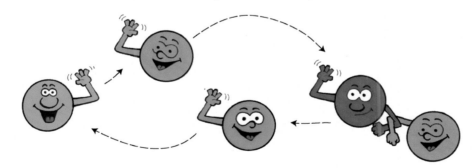

3 Two new molecules are made.

SODIUM CHLORIDE
(table salt)

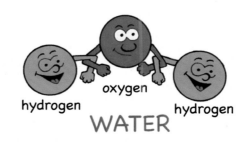

WATER

5.3 Molecules Join Together

Sometimes when molecules meet, they join together. In this example, two chlorine atoms in a chlorine gas molecule meet two sodium atoms. The chlorine atoms and the sodium atoms combine to make table salt!

1 Two chlorine atoms [in a chlorine gas molecule] meeting two sodium atoms.

chlorine   chlorine   sodium   sodium

CHLORINE GAS MOLECULE   SODIUM ATOMS

2 The chlorine atoms join the sodium atoms to make two sodium chloride molecules [table salt].

sodium   chlorine   sodium   chlorine

TABLE SALT

5.4 Molecules Break Apart

Sometimes molecules simply break apart to form new molecules. In the illustration below, two water molecules break apart, and then the atoms join together in a different way to make oxygen gas and hydrogen gas.

1 Two water molecules.

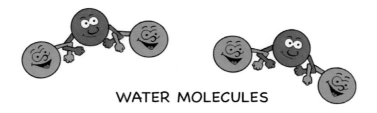

WATER MOLECULES

2 The water molecules break apart.

3 New molecules are made.

HYDROGEN GAS   OXYGEN GAS

All of these examples show how molecules and atoms react with one another. It is important to realize that in every reaction atoms are neither created nor destroyed. Atoms can rearrange themselves by combining in different ways and changing places to make new molecules, but atoms never simply appear or disappear.

5.5 Reactions Are Everywhere

When atoms join together, leave a molecule, or switch places in molecules, a chemical reaction has occurred.

There are lots of chemical reactions. They go on all the time and all around us. For example, the gasoline inside a car reacts with oxygen to provide energy for the car to move.

Reactions occur when you bake bread or cook an egg.

When you leave your metal shovel out in the rain, the red rust that forms on it is caused by a chemical reaction.

Reactions occur inside your body. When you eat a piece of cheese or drink a glass of milk, reactions occur inside your mouth. These reactions help break down the food

molecules into smaller pieces. Inside your stomach there are strong molecules that break your food down still further.

Even when you breathe, reactions inside your lungs help oxygen get into your blood so it can be carried through your body. Reactions are everywhere.

5.6 Reactions Follow Rules

Reactions also have to follow rules. Not every molecule will react with every other molecule or atom. Some molecules won't react at all. For example, the noble gases, such as neon, helium, and argon, usually don't react with any other molecules.

Some molecules react with lots of other molecules. Water will react with many different atoms and molecules. Water will even start a fire during some reactions!

5.7 We Can See Reactions

Often we can observe something happening if a reaction is occurring. Sometimes we can see bubbles. Sometimes we might see little particles form that look like sand. Sometimes the glass might change temperature in our hands if a reaction is happening. There could also be fire, an explosion, or a color change.

All of these observations tell us a reaction may be happening.

Bubbles

Fire

Particles
(like sand)

5.8 Summary

○ When atoms get rearranged a chemical reaction has occurred.

○ Atoms can switch places, join together, or separate from each other during a chemical reaction.

○ In a chemical reaction atoms rearrange but are never created or destroyed.

○ Reactions occur everywhere.

○ We may be seeing a reaction take place when there are bubbles, color changes, or temperature changes.

Chapter 6 What Is Biology?

Biology

6.1 Introduction

In Chapter 1 you learned that science is a way to study how things work, and in Chapters 2-5 you learned about the first building block of science—chemistry.

In this chapter we will take a look at the second building block of science called biology. Biology is the study of life. Biologists look at how plants and animals live, grow, and interact with each other.

6.2 History of Biology

People have been studying life for thousands of years. The very first biologist is said to be Aristotle. Aristotle lived in Greece from 384–322 B.C.E. Aristotle thought about the world around him, and he was particularly interested in living things.

ARISTOTLE 384–322 BCE

But Aristotle didn't just think about living things, he also made observations. Making an observation happens when you examine something with your eyes, or touch something with your fingers, or smell something with your nose.

Aristotle made many observations of both plants and animals. He looked at the parts of animals, how animals move, and how animals breathe or sleep. He also observed how some animals are similar to each other and how some animals are different.

Aristotle had many ideas about life and he wrote many books to describe his ideas to others. In the same way, modern biologists observe life and write their ideas in books and papers so they can share those ideas with other biologists and other scientists.

6.3 Modern Biology

Today, modern biologists continue to observe life like Aristotle did thousands of years ago. However, unlike Aristotle, modern biologists can use chemistry and physics to help understand how living things work.

For example, modern biologists can study how plants make food from light by studying the chemicals in plant leaves and the physics of light waves.

Modern biologists have many tools that Aristotle didn't have. With these tools it is possible for modern biologists to make many more observations about life than Aristotle was able to.

For example, modern biologists can use microscopes to see life that is too tiny to see with our eyes.

And modern biologists can also use airplanes to observe how animals move in large groups.

6.4 Everyday Biology

Learning about life is something everyone does.

If you care for a pet or observe an ant carry away your picnic food, you are doing biology.

When you plant a garden and observe how your vegetables grow, you are doing biology.

If you make yogurt or watch bread rise, you are doing biology.

Biology is simply observing, studying, and working with living things.

Biology can involve very complicated experiments or be as simple as observing a spider trap a moth in a web. The more observations you make and the more details you observe, the more you will discover about biology!

6.5 Summary

- Biology is the study of life.

- Aristotle is considered the first biologist.

- Modern biologists use chemistry and physics to understand living things.

- Biology is observing, studying, and working with living things.

Chapter 7 Life

Biology

7.1 Studying Life

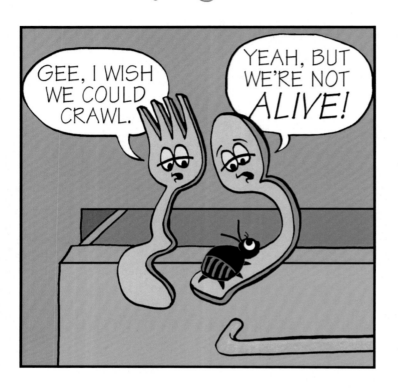

What makes plants, dogs, and beetles different from rocks, dirt, and metal? Maybe you have noticed that rocks don't move like dogs and dirt doesn't need food. Maybe you have seen that forks and knives, which are made of metals, don't crawl around in the kitchen like beetles do. Living things are different from rocks and dirt and metals because living things are alive.

What does it mean to be alive? Think about how you are different from a rock. You need food and a rock doesn't. One feature of being alive is needing food.

Second, you can walk, run, jump, curl up into a ball, and roll on the carpet. But a rock can't move. So another feature of being alive is the ability to move.

Finally, a rock can't make baby rocks, but plants, animals, and humans all make baby plants or baby animals or baby humans. So another feature of being alive is the ability to reproduce.

As we can see, living things are much different from nonliving things.

7.2 Sorting Living Things

How do we keep track of all of the living things we find on the planet? Is there a way to sort them? Why would we want to sort them?

Sorting living things helps us understand how they are different and how they are the same. For example, what if you had some yellow blocks and some blue blocks. How would you sort them?

HOW WOULD YOU SORT THE BLOCKS?

If you sort your blocks according to color, you can see that the blue blocks are different from the yellow blocks. However, you might also notice that some of the blue blocks are the same size as some of the yellow blocks.

So you could also sort them by size. By sorting, you can see how some blocks are different (different color), but also how some are the same (same size).

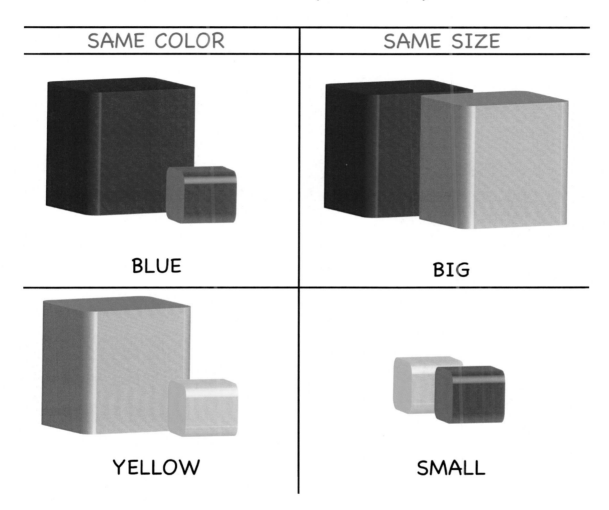

SAME COLOR	SAME SIZE
BLUE	BIG
YELLOW	SMALL

A very long time ago a man named Carolus Linneaus thought about how to sort living things. He came up with a system of sorting all of the creatures on the planet. We call this system taxonomy.

Taxonomy is a branch of biology that is concerned with how to sort living things.

We sort creatures by looking at different features.

A feature is anything like hair, hooves, feathers, or green leaves.

For example, we might sort animals that have hair from animals that don't have hair. We might sort plants that live in the soil from plants that live in the water.

We might also sort very small creatures that we can't even see with our eyes from larger creatures that we can see. Looking at the features of living things helps us sort them.

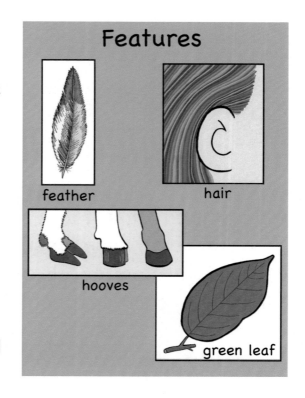

7.3 Domains and Kingdoms

It is very difficult to decide how to sort living things because they have so many different features! Once upon a time, living things were sorted into only two large groups—plants and animals. However, as scientists learned more about all of the different creatures, they had to make more groups.

House plants are in the kingdom Plantae

Today, scientists use three large groups to sort living things. These groups are called domains. The names of the domains are Eukarya, Bacteria, and Archaea. These domains are then further divided into six kingdoms which are Plantae, Animalia, Protista, Fungi, Bacteria, and Archaea.

The kingdom Plantae groups all of the plants. Houseplants are in the kingdom Plantae.

Dogs are in the kingdom
Animalia

Animalia groups all of the animals. Dogs are in the kingdom Animalia. So are cats, frogs, and butterflies.

Bacteria and Archaea are kingdoms that group some of the very small creatures that we can't see when using only our eyes.

Protista is a group for very small creatures called protists.

And finally, Fungi groups things like toadstools and mushrooms.

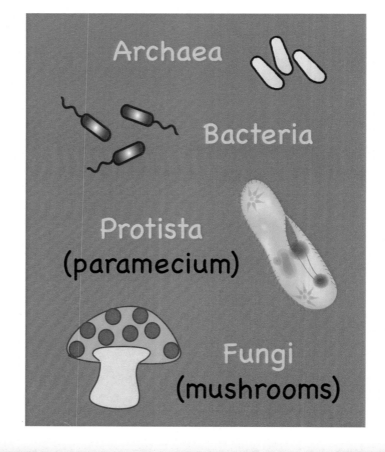

7.4 Sorting Within Kingdoms

Scientists first sort living things into their domains and then sort them into their kingdoms. Then scientists organize the living things into smaller groups to better understand them. So, living things in different domains are sorted into kingdoms and then further sorted into smaller, different groups.

To sort living things into smaller groups, scientists again look for different or similar features. For example, both birds and cats are animals, but we can see that birds are different from cats. For one thing, birds have wings and fly, but cats don't fly. Cats have fur and eat birds.

Even though birds and cats are both animals, they are different from each other. All of the birds are put into a group for birds, and all of the cats are put into a group for cats.

What about tigers and house cats? They are both cats. Are they exactly the same? In fact, they aren't. Even

though tigers and house cats have some similar features, they are also different. For example, house cats don't usually eat their owners, but tigers could! So house cats and tigers are put into even smaller groups within the larger grouping of "cats."

7.5 Naming

How do we name all of the creatures we find? Because there are so many different languages and so many different living things, scientists use a scientific name to name each living thing. Every plant and animal, fungus, bacterium, and archaeon has a scientific name. The scientific name for each living thing comes from the Latin language. Each creature has two Latin names. The first name is called the genus, and the second name is called the species.

The Latin name for a house cat is *Felis catus,* and the Latin name for humans is *Homo sapiens,* which means "man wise."

7.6 Summary

○   Living things are different from nonliving things. All living things need food and can reproduce, and some living things can move.

○   Scientists sort living things into groups to understand them better.

○   Domains and kingdoms are two kinds of groups that scientists use to sort living things.

○   All living things have a special scientific name which is in Latin.

Chapter 8   The Cell:
A Tiny City

Biology

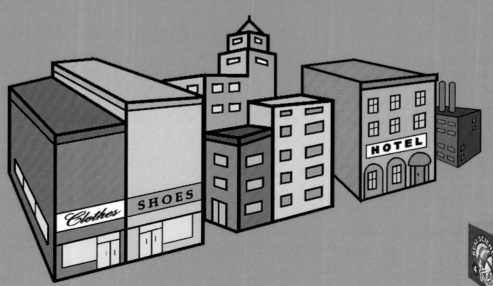

8.1 Creatures

Have you ever wondered what makes a frog a frog or a rose a rose? Have you ever wondered what you are made of?

We saw in Chapter 7 that living things are different from nonliving things. But what are living things made of?

As we saw in Chapters 2-5 everything we can touch, feel, and smell is made of atoms. Even living things are made of atoms. In fact, both rocks and frogs are made of atoms.

We also know from chemistry, that atoms fit together to make molecules. Because molecules are different and fit together in different ways, rocks and frogs are different. In this chapter we will explore how molecules fit together to make living things like frogs different from nonliving things like rocks.

8.2 The Cell

The atoms and molecules in living things are designed to fit together in special ways to make cells.

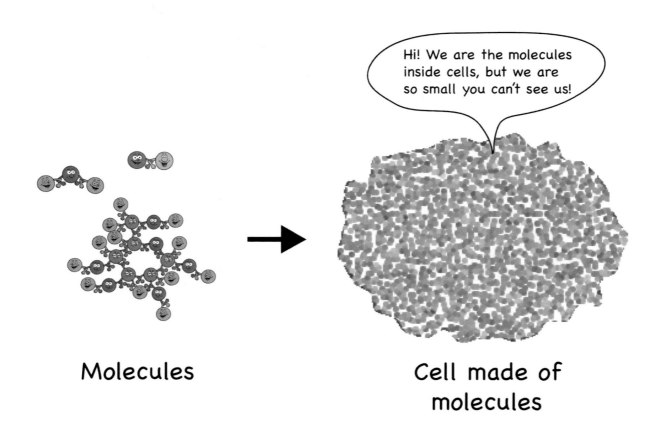

Hi! We are the molecules inside cells, but we are so small you can't see us!

Molecules

Cell made of molecules

The atoms and molecules inside cells are very small, so you can't see them. But when atoms and molecules are put together, they make cells.

Cells are designed to fit together in special ways to make all of the parts for living things. For example, molecules fit together to make skin cells and skin cells fit together to make skin.

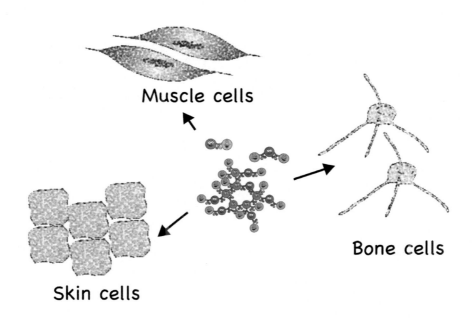

Muscle cells

Bone cells

Skin cells

Molecules also fit together to make muscle cells and muscle cells fit together to make muscles.

Bone cells, muscle cells, and skin cells fit together to make arms or legs or fingers.

8.3 A Tiny City

Cells are like tiny cities. Imagine what it takes to make even one item we use in a city—like milk. What does it take to make milk?

We know that milk comes from cows. Cows are not usually in the city, but in the country, so we have to go to the country to get the milk from the cows.

Once we have the milk, we have to move it to the city. We need a big truck to move the milk.

But it can't be any kind of truck, it has to be a special dairy truck. If it were a sand truck we might get sand in our milk. We must use a milk truck.

Once the milk is in the city, what happens to it? It would be hard to get our milk directly from the milk truck so the milk is put into cartons.

Once the milk is put into cartons we have to put it somewhere we can find it, like a grocery store. Once it is in the grocery store, your mom or dad can buy it and put it on the table!

Think about all of the people it takes to bring milk to your table. We need people to milk the cows, clean the raw milk so we can drink it, move the milk to the truck, drive the truck into town, put the milk into cartons, and finally put the milk in the grocery store so your mom and dad can buy the milk.

It takes a lot of people to perform all those tasks!

Also, all the tasks have to all be done in the right order. Imagine what would happen if the milk was put into cartons in the wrong order. What would happen if the cartons were sealed before the milk was put in?

The process for making milk is similar to how a cell works. There are lots of complicated molecules called proteins doing all kinds of jobs inside a cell that keep the cell alive.

Each job has to be done by a special protein and all the jobs have to be done in a particular way and in a particular order.

8.4 Parts of a Cell

Cells have lots of different parts like cities have lots of different places. We have talked about how cities provide milk in a grocery store, but there are also other places in cities that provide other things.

For example, there is usually a downtown area in a city that has many important places, like the courthouse or the tax department. Scattered through the city are post offices where letters are mailed back and forth, grocery stores for buying milk, hotels for people who visit, and factories for making things.

There are lots of different places and lots of people doing lots of different jobs.

Nucleus (building)

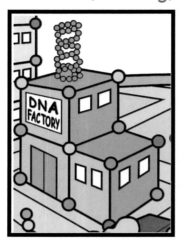

Cells are like cities in this way. In most cells there is a central part that does many of the important jobs in the cell. This central part is called the nucleus. The nucleus of human and plant cells holds all of the important information, called DNA, for the cell.

Golgi Apparatus (building)
Ribosomes (protein man)

Outside the nucleus there are lots of places that do special jobs inside the cell. There is a place called the golgi apparatus which makes proteins. There are also special proteins called ribosomes that make other proteins.

Kinesin (truck)
Microtubules (road)

There are "roads," called microtubules, that move proteins from place to place. And there are "trucks" called kinesin that carry the proteins on the "roads."

A City Cell

Putting it all together, the cell looks a lot like a little city with every protein and molecule doing their part!

8.5 Summary

- Atoms and molecules fit together to make cells.

- Cells fit together in special ways to make the parts for living things.

- All the parts of a cell have special jobs.

Chapter 9 Viruses, Bacteria, and Archaea

Biology

9.1 Introduction

Now that we know that living things are made of atoms and molecules that fit together to make cells, let's take a look at some of the smallest living things.

You might think that the smallest form of life is a mouse or a plant. But there are even smaller forms of life that you can't see with your eyes alone.

Have you ever wondered what happens when you catch a cold or get an infection after you get a cut? Colds and infections are caused by two of the smallest forms of life—viruses and bacteria.

Archaea are another very tiny form of life. Viruses, bacteria, and archaea can only be seen with a microscope.

9.2 Viruses

Viruses are very simple and very small. Viruses don't even have all the parts of a complete cell. Viruses are just a bag of molecules covered with more molecules. Because viruses are not true cells, they must live inside other organisms (living things) to survive.

Some scientists are debating about whether or not viruses can really be called "alive." Viruses are usually classified according to the type of organism they are living inside, and viruses don't have their own kingdom.

Viruses cause diseases in plants, animals, and people.

The common cold virus is very small and looks like a soccer ball with spikes. This virus can live for many days on nonliving objects, like door handles or pencils. When you touch an object that has a cold virus on it, the virus can enter your body, especially if you touch your mouth or face. If the virus has time to settle in, it will start to multiply in your body and make you sick. Coughing and sneezing are ways that your body uses to try to get rid of the virus.

The best way to keep from catching a cold virus is to wash your hands often!

9.3 Bacteria

Bacteria are larger than viruses. Bacteria have all the parts of true cells and are considered by all scientists to be alive. Bacteria eat, build molecules, grow, and divide to make new bacteria. Some bacteria can even move on their own.

Bacteria have unique shapes. Some are round, like balls. Some are rod-shaped, like hot dogs. Some are spiral shaped, like a spring.

Some bacteria stay by themselves. Other bacteria like to stick to each other to form colonies. A colony of bacteria is just a large group of individual bacteria that are all living very close together in a group.

WE LIKE DIRT.

Bacteria live everywhere. Some bacteria live in the dirt and help plants get food from the soil. Without bacteria, plants could not get the food they need to grow.

Bacteria can also live in water. Sometimes the bacteria in water are food for other living things. But sometimes the bacteria in water can make us sick. One reason we clean our drinking water very carefully is to remove the bacteria that could make us sick.

Bacteria also live on us and inside of us. Bacteria live on our skin, in our mouths, throats, stomachs, and intestines. Some bacteria help us digest our food.

When bacteria live under the right conditions, they will divide and make more bacteria. In order to stay alive, all bacteria need to have enough water available. However, some bacteria can survive even when there is a lack of water, but they will not be able to grow and divide until more water becomes available.

9.4 Archaea

Archaea are similar to bacteria but are different enough to be classified in a group of their own.

Archaea have cell types that are slightly different from those of bacteria. Archaea are often found in places that have very extreme conditions, like places that are very hot or very salty. Some archaea even eat sulfur or iron for food!

9.5 Summary

○ Viruses, bacteria, and archaea are tiny organisms.

○ Viruses, bacteria, and archaea are able to live under many different conditions.

○ Viruses typically cause illness.

○ Some bacteria are helpful. Other bacteria can cause illness.

○ Archaea are both similar to and different from bacteria.

Chapter 10 What Is Physics?

10.1 Introduction

In the previous chapters we explored the first two building blocks of science—chemistry and biology. In this chapter we will take a look at the third building block of science: physics.

Have you ever thrown a ball up in the air? Did you notice the ball when it left your hand? What did it do? Did it go up? Did it come back down? Unless it gets stuck in a tree or picked up by a big bird, a ball that is thrown up into the air will always come back down.

Have you ever tried to throw a ball really far or really high? Have you ever watched how far or how high the ball goes? Have you ever noticed that it's harder to throw a heavy ball than it is to throw a light ball? Have you ever noticed that it's almost impossible to throw a feather?

Physics is the branch of science that explores how far or how high a ball might go or how heavy it needs to be so that it can be thrown. Scientists who study physics are called physicists.

10.2 History of Physics

Physics is about studying the way things behave and then figuring out the rules those things follow to make them behave that way. Physicists don't make the rules, but they discover the rules by watching how the world works.

Aristotle studied motion, but it was Galileo Galilei, an Italian astronomer, who used physics to understand how things move.

Galileo is known for a famous experiment where he dropped two balls off a building to see what would happen. He used a heavy ball and a light ball. To everyone's surprise, he found out that they both hit the ground at the same time!

Physicists also use math to figure out the rules. Isaac Newton was a great scientist and mathematician who figured out many important rules of physics. By using math, Newton figured out exactly why the balls Galileo dropped hit the ground at the same time. Math is an essential part of physics and helps us understand the rules of physics.

10.3 Modern Physics

Did you know that balls will follow the same rules no matter where you are on the Earth? You can be in the frozen Arctic, and if you drop two balls, they will fall in exactly the same way. You can be in a desert, at the beach, or on a boat, and if you drop two  balls, they will fall in exactly the same way. No matter where you are on Earth, a ball will always follow the rules of physics!

How balls behave when dropped is explained by the rule "what goes up – comes down" which is a rule about gravity. Gravity is what makes the balls come back down.

Gravity is also what keeps you from flying off the surface of the Earth. Balls, toys, cars, houses, and even birds obey the rules of gravity.

10.4 Everyday Physics

Every day, physics is happening all around you. In the same way that you learn chemistry and biology, to learn physics you need to make observations. When you are looking at something with your eyes, you are making an observation. Making good observations is the first and most important step when you are trying to understand physics.

When you make an observation, try to notice everything you can about what you are observing. If you are at the movie theater getting popcorn, try noticing the popcorn machine. Where does the popcorn go in? Where does it come out? What is moving on the machine? What is staying still?

Notice the popcorn when it comes out. Is it hot or is it cold? These kinds of observations are important if you want to think like a scientist.

10.5 Summary

- Physics is about studying the way things behave and then figuring out the rules those things follow that make them behave that way.

- Objects, like balls, planes, and birds, always obey the rules of physics.

- The rules of physics are true no matter where you are on the Earth.

- Physicists don't make the rules, they discover the rules.

- To think like a scientist, you must make good observations.

Chapter 11 Push and Pull

Physics

11.1 Up the Hill

What do you think would happen if you tried to take your baby sister up a hill in a wagon? You might start on flat ground at the bottom of the hill. First, you might grab the handle of the wagon, and as you did so, you might feel the wagon pull against you. You might think that the wagon with your sister in it is heavy.

Then, to get the wagon moving by pulling on it, you might need to use all of the strength in your arms and legs.

Once the wagon is moving on flat ground, you might find that it is easy to roll your sister along and that you don't have to use as many muscles. But, as you get to the bottom of the hill, you might need to use all your muscles again to pull the wagon uphill.

When you reach the top of the hill, you might discover that you are completely out of breath and a little tired. You might say that you used all your energy to do the work you needed to do to get your sister up the hill. And a physicist would say you are exactly right!

When you pull your little sister up a hill in a wagon, you are doing work and you are using energy. In physics, work is what happens when force moves an object. Energy gives you the ability to do work. But what is force? And what is energy?

11.2 Force

When you are pulling the wagon up the hill, you are using a force. Force is any action that changes the location of an object. Because you are changing the location of the wagon (and your sister) you are using force.

Force is also any action that changes the shape of an object. If you were to squeeze a marshmallow, you would be using force. By squeezing a marshmallow, you are changing its shape with the force created by your hands.

Force is also any action that changes how fast or how slowly an object is moving. You may have experienced this kind of force if you ever tried to catch the same ball as your friend. If you were both looking at the ball and not where you were going, you might have run into each other. When that happens—WHAM!—the two of you collide, and you both stop moving. In this action you each used force to stop the other from moving. You could probably feel the effects of the force stay with your head or knees for several hours!

Force is any action that changes the location of an object, the shape of an object, or how fast or slowly an object is moving.

11.3 Work

In the first section we saw that work happens when force moves an object. Work also happens when force changes the shape of an object or when force changes how fast or how slowly an object is moving.

How is work related to force? If you use more force are you doing more work? Maybe. Or if you are doing more work, are you using more force? Yes!

Imagine that instead of one little sister, you have two. And imagine trying to pull both little sisters up the hill in the wagon.

It will take more muscles, more energy, and more effort to move two little sisters up the hill. In fact, if your little sisters were twins and weighed exactly the same amount, you would have to do twice the amount of work to get both of them up the hill.

The same is true for squeezing a marshmallow or running into your friend. If you squeeze the marshmallow more, you are doing more work. If you and your friend are running faster and generating more force, you are doing more work when you run into each other.

11.4 Energy

How do you get the energy for pulling your little sister up the hill in a wagon or for squeezing a marshmallow or for colliding with your friend? Where does the energy come from?

The energy for you to do all these things comes from your breakfast. When you eat breakfast, you are giving your body the energy it needs to do work. In physics, energy is something that gives something else the ability to do work. When you eat eggs or toast or

cereal, your body takes the energy in the food and uses it in a way that helps your muscles move.

The food you eat for breakfast gives your muscles the ability to do work—like pulling a wagon full of sisters up a hill! And that's a lot of work.

11.5 Summary

- A force is any action that changes the location of an object or the shape of an object or how fast or slowly an object is moving.

- Work is what happens when a force moves an object.

- Energy is something that gives something else the ability to do work.

Chapter 12   Types of Energy

Physics

12.1 Stored Energy

Recall from the last chapter that energy is something that gives something else the ability to do work. We saw how the food you eat for breakfast gives you the ability to move your muscles, which allows you to pull your little sister up a hill.

The food you eat for breakfast is a type of stored energy. Stored energy is energy that has not been used. A box of cereal may not look like it is full of energy, but in fact it has lots of energy molecules.

These energy molecules are called carbohydrates. When your body needs energy, it will break down the carbohydrate molecules so they can be used for moving muscles or walking to the store.

12.2 Types of Stored Energy

There are different types of stored energy. Have you ever taken a rubber band and stretched it across your finger and thumb? What happens when you release your thumb, letting the rubber band go? It flies through the air. There is stored energy in a rubber band.

But can you use this stored energy for breakfast? NO! You wouldn't eat rubber bands for breakfast!

A rubber band has a different type of stored energy than your breakfast cereal.

Breakfast cereal has chemical stored energy. Chemical stored energy is energy that comes from chemicals and chemical reactions.

A rubber band has elastic stored energy. Elastic stored energy is energy that is found in materials that can stretch.

A book on a table has gravitational stored energy. It could do work if it were to fall to the ground and smash an egg. Gravitational stored energy comes from objects that are elevated above the ground and can be pulled down by gravity.

12.3 Releasing Stored Energy

When you let a rubber band go from your thumb, you release the stored elastic energy and the rubber band flew through the air. When you eat cereal for breakfast, your body breaks down the carbohydrate molecules, releasing the stored chemical energy so your muscles can pull a wagon. If a book falls off a table and onto the floor, the stored gravitational energy of the book is released and can be used to break an egg.

When each of these types of stored energy is released, the energy does not disappear but is converted into moving energy. The rubber band moved, the wagon moved, and the book moved.

In order for stored energy to do work, it must first be released. The stored energy in the rubber band, the breakfast cereal, and the book were all released and changed into moving energy. But what is moving energy?

12.4 Moving Energy

Moving energy is the energy found in moving objects. There is only one type of moving energy, and physicists call this energy kinetic energy.

Kinetic energy can come from different types of stored energy, but only the energy of a moving object is called kinetic energy.

When you release the stored energy in a rubber band and it's sitting on the floor, it can no longer do any work unless it is picked up and stretched again. When the rubber band is sitting on the floor, all of the stored energy has been released, and all of the kinetic energy has been used. You might think that the energy is lost. However, none of the energy disappears, it just gets changed to a different type of energy.

12.5 Summary

- Stored energy is energy that has not been used yet.

- There are different types of stored energy. Breakfast cereal has chemical stored energy. A rubber band has elastic stored energy.

- When stored energy is released, it can be changed into kinetic energy—the energy of an object that is moving.

Chapter 13   Saving Energy

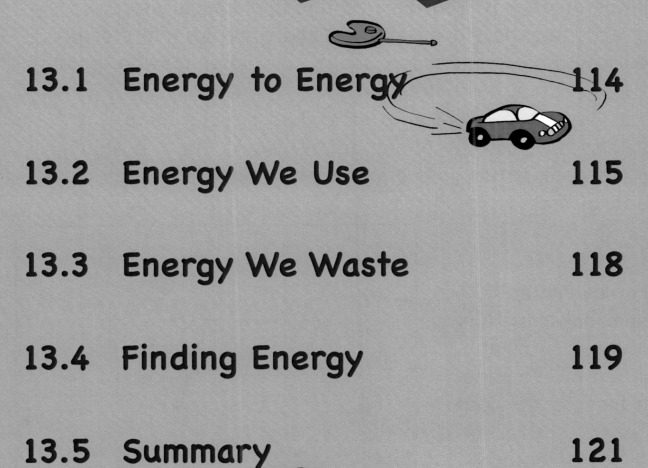

Physics

13.1 Energy to Energy

In the last few chapters you learned how energy is used when a force does work. You also learned about different kinds of energy. You learned about chemical stored energy, elastic stored energy, and gravitational stored energy. You also learned about kinetic, or moving, energy.

You learned that when a stretched rubber band is released, it can no longer do any work. But the energy is not lost, it has just changed to a different type of energy.

For example, if you look carefully at your brother when he is riding a bicycle, you can observe different types of energy being used to generate different kinds of forces and work.

Your brother's body is using chemical energy from his breakfast to move his muscles. The muscles are using the chemical energy to move the pedals on the bike. The pedals on the bike are connected to a chain and a gear. As the chain and the gear move, the wheels move. The chain, gear, and wheels are all using mechanical energy to move the bike forward.

In this example, you can see how one type of energy is getting converted, or changed, into another type of energy. As we saw in Chapter 12, energy is only converted (or changed) into other types of energy. Energy is never created, and energy is never destroyed. It is only changed from one form to another.

13.2 Energy We Use

When your dad puts gas in the car, he is giving the car energy it can use. Your body can't use gasoline to move the pedals on a bicycle, but a car uses gasoline to run the motor. Likewise, a car cannot use cereal to move its motor like you use cereal to give energy to your legs.

There are different forms of energy, and not all types of energy can be used in the same way.

Gasoline is one form of energy that is used to power things like cars and boats. Cereal and bread are another form of energy used to power things like human bodies.

Batteries are yet another form of energy used to power things like flashlights and laptop computers.

When we "use" energy, we are converting one form of energy to another form of energy.

When you use batteries in a music player, the chemical energy in the batteries is converted to moving and sound energy in the music player.

Eventually the batteries run out. All the chemical energy is converted to moving and sound energy, and there is no more chemical energy in the batteries. This is how we "use" energy. We convert it from one form to another form.

13.3 Energy We Waste

Even though energy cannot be destroyed, it is possible to "waste" energy. You might have heard your dad telling you to turn off the lights after you leave a room. Or you might hear your mom tell you not to leave the door open in the middle of winter. They may have told you not to "waste" energy. But if energy cannot be destroyed, what does it mean to waste energy?

Energy is "wasted" when energy is excessively or unnecessarily converted from one form to another form.

If you are playing with your battery powered car, then you are converting one form of energy (chemical energy in the battery) to another form of energy (moving energy in the car). Because you are playing with the car, it is necessary to convert the chemical energy.

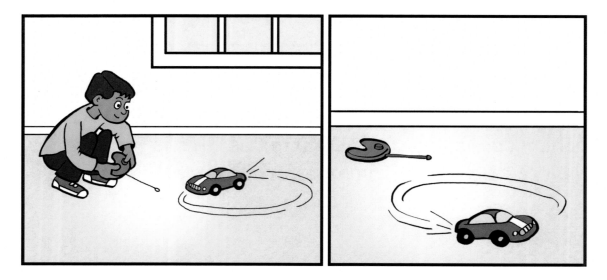

But if you walk away from the car, and you forget to turn it off, the battery is running but you are no longer playing with the car. Now you are "wasting" the energy in the battery because you are not using it to play with the car. When you go back to your car the next day, you discover it won't work anymore because the battery is "dead." All the chemical energy in the battery is gone. This is wasting energy. You didn't destroy the energy, you just unnecessarily converted it from chemical to moving energy when you didn't need to.

13.4 Finding Energy

We get much of our energy from the Earth. Gasoline, coal, and natural gas all come from inside the Earth. Nuclear energy also comes from the Earth in the form of plutonium. It is possible to get energy from moving water or wind, and we can also get energy from the Sun.

The food we eat comes from the Earth. We grow plants and raise animals to get food for our bodies. The plants get their energy from the Earth and also from the Sun.

Will we ever run out of energy? Yes and no. The energy we get from gasoline comes from fossils. There are only so many fossils in the ground, so it is possible that one day we will have used all the energy that is stored in fossils. When that happens, we will not have any more gasoline that comes from fossils to run our cars. This is also true with other forms of energy that come from fossils, such as coal or natural gas.

But if we remember that energy cannot be destroyed, only converted from one form to another, it may be possible to discover new ways to convert energy. Maybe there are new ways to convert the Sun's energy to chemical energy. Or maybe there are ways to get chemical energy from rocks. Or maybe there are ways to get electrical energy from grass or trees. Maybe you will be the next scientist who discovers a new way to convert energy into a form that can be used to power cars or boats or planes!

13.5 Summary

- Energy is neither created nor destroyed.

- Energy is converted from one form to another form.

- Energy is wasted by converting it from one form to another form excessively or unnecessarily.

- There may be new ways to convert energy to a form that can be used for fuel.

Chapter 14 What Is Geology?

14.1 Introduction

We have explored three important building blocks for science—chemistry, biology, and physics.

We are now ready to take a look at a fourth building block of science: geology.

Geology is the study of the Earth. By studying the Earth, scientists attempt to understand what the Earth is made of, how Earth came into being, how Earth has changed in the past, how it is changing now, and our role as we live on Earth.

Do you ever pick up rocks and wonder how they were made and what they are made of? Do you sometimes look at mountains and wonder how they were formed? Have you wondered what's at the bottom of the ocean? Have you noticed how weather affects the landscape? Do you wonder why certain birds and wild animals live near you and others don't? These are all questions

that are explored by scientists who study geology.

14.2 History of Geology

Depending on where you live, when you go outside and walk around, you'll see many different features of Earth. You might see mountains or rivers. You might see fields of dirt or fields of grass. You might see lakes or oceans, mesas or glaciers, forests or prairies.

Ancient people also saw many of the same features you see. Although lakes come and go and rivers might change course, many of the features you see today are the same features ancient people would have seen.

One of the very first people to study Earth's features was the Greek philosopher Theophrastus who lived from about 371-287 BCE.

Theophrastus was a student of Aristotle. Recall from Chapter 6 that Aristotle was a Greek philosopher who was the first to study plants and animals. Like Aristotle, Theophrastus was interested in science. He studied rocks and explored what happened when rocks were heated.

THEOPHRASTUS
371-287 BCE

Many of the first geologists also asked questions about how the Earth came into being and how many years the Earth has existed. All of these questions shaped the modern science we now call geology.

14.3 Modern Geology

Modern geologists continue to study rocks and what rocks are made of. They also ask questions about how mountains, rivers, and glaciers form. Modern geologists have an advantage over ancient geologists because modern geologists can use chemistry and physics to better understand how things work.

There are different kinds of modern geologists. Some modern geologists focus on the chemistry of Earth. These geologists are called geochemists. Geochemists study how atoms and molecules make rocks, soils, minerals, and fuels.

Other modern geologists focus on the structure of Earth. These geologists are called structural geologists. Structural geologists study how Earth is put together and how it changes. They are interested in how rocks change and what makes mountains and valleys.

There are also modern geologists who study how humans affect the water, air, and land quality of Earth. These geologists study Earth's environment and are called environmental geologists.

14.4 Everyday Geology

Even though you may not be a geologist yet, you can learn about the Earth by simply observing what happens around you.

What happens when it rains? Do the roads flood? Do you get mud slides, or does a river find a new path? What happens in the hot sun? Do you observe mud forming cracks or rocks crumbling? Have you ever been in an earthquake? Did you feel the ground move?

Paying attention to where you live, what happens during storms, and how the land around you changes over time are activities you can do every day.

14.5 Summary

- Geology is the study of Earth.

- The first geologists looked at rocks and minerals and asked questions about how Earth came into being.

- Geochemists are modern geologists who study how atoms and molecules form Earth.

- Structural geologists look at how Earth is put together.

- Environmental geologists look at changes in the quality of the water, the air, and the land on Earth.

Chapter 15: What Is Earth Made Of?

Geology

15.1 Introduction

If you walk outside, do you notice where you live? If you live in the country, do you notice the trees and grass? If you live in a city, do you notice the buildings and streets?

If you look down, do you notice the ground? If you live in a city, do you notice how much of the ground is covered up with streets or pavement? If you live in the country, do you notice if the ground has grass, rocks, or dirt?

When you notice the ground with the rocks, dirt, grass, and trees, you are noticing the Earth. The Earth is where you live.

But what is the Earth made of? What is dirt? What are rocks? Why do grass and trees grow in dirt? How deep does the dirt go? How many kinds of rocks are there? What is below the rocks and dirt? More rocks? Trees? Chocolate syrup?

In this chapter we will learn about what the Earth is made of.

15.2 Rocks and Minerals

The crust is the outer part of Earth and is where we live. The crust is mostly made of rock. If you go outside and start digging in the ground with a shovel, you will eventually hit some type of rock.

What is the difference between a rock and a tree? Trees are living things. Rocks are not living things. Rocks do not move like living things. Rocks do not grow like living things, and rocks do not multiply like living things.

Living things are made mostly of carbon, but rocks are made mostly of silicon. Silicon and carbon are atoms (also called elements). Because rocks are made mostly from silicon and living things are made mostly from carbon, rocks are very different from living things.

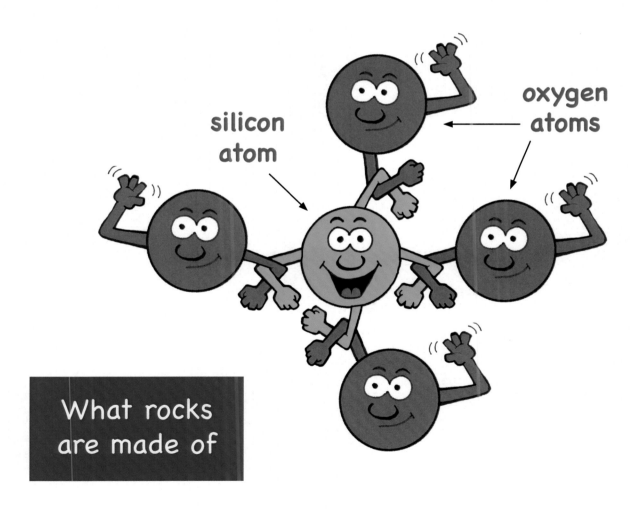

silicon atom

oxygen atoms

What rocks are made of

All rocks come from magma, which is molten (melted) rock deep inside the Earth. Magma is made mostly of the elements silicon and oxygen. Rocks form when the magma cools and mixes with other elements, like magnesium, iron, or aluminum.

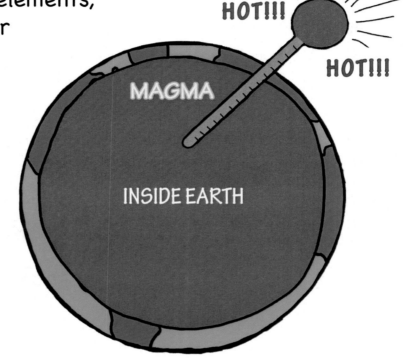

When magma cools very slowly, the atoms in the magma have a chance to line up in an orderly fashion. When this happens, the material that is formed is called a mineral.

There are many different kinds of minerals. One common mineral found in rocks is quartz. Quartz can be clear, pink, purple, or other colors.

Mica is also a mineral. Mica is soft and looks like thin, layered paper. You can peel mica sheets away from each other.

Calcite is another mineral found in rocks. Calcite is made of calcium and oxygen and forms beautiful crystals that come in different colors.

Minerals

Quartz   Mica   Calcite

Some minerals are used in jewelry. Rubies are a brilliant red-colored mineral made of aluminum and oxygen. Emeralds are a different type of mineral made of aluminum, beryllium, silicon, and oxygen. Emeralds have a deep green color. Both rubies and emeralds are hard to find, and that is why jewelry made with them is often expensive.

Ruby Crystal

There are also many different kinds of rocks. There are rocks that are formed deep inside the Earth, and there are rocks that form on the surface of the Earth. There are also rocks that have been changed from one type of rock into another type of rock. The different types of rocks are called igneous, sedimentary, and metamorphic. Minerals are the building blocks of all these types of rocks.

Types of Rock

Igneous
(granite)

Sedimentary
(limestone)

Metamorphic
(garnet schist)

Igneous rocks are the most plentiful type of rock. Igneous rocks are formed when molten magma inside the Earth cools and hardens. Granite is one kind of igneous rock, and it has a lot of quartz in it.

Sedimentary rocks are formed from bits of rocks and other materials left behind by wind or water. As these materials pile on top of each other, layers form. These layers are pressed tightly together, turning the materials into rock.

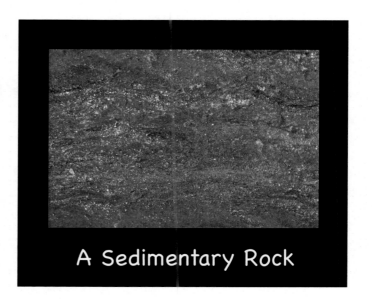

A Sedimentary Rock

A Metamorphic Rock

Metamorphic rocks are those that have changed from one type of rock into another. Very high heat and pressure cause these changes. Igneous rocks and sedimentary rocks can be changed into metamorphic rocks. Even metamorphic rocks can be changed into other metamorphic rocks!

15.3 Dirt

When you go outside and dig a hole in the ground, you not only find rocks, but you also find dirt. Dirt is made of rocks, plants, minerals, and even small animals!

There are many different types of dirt. Some dirt is sandy and light in color. Other dirt is moist and dark in color. Have you ever wondered why some parts of the world are used for growing food and other parts of the world are not? That's because some dirt is good for growing plants and other dirt is not.

Dirt is also called soil. Dirt and soil come mainly from rocks. Because different parts of the Earth have different kinds of rocks, there are different types of soil in different places.

15.4 Summary

○ The Earth is made mostly of rock.

○ All rocks come from magma.

○ Minerals form when magma cools slowly, allowing the atoms to line up in an orderly fashion.

○ Dirt is called soil. It is made mostly from rocks and contains materials that come from plants, minerals, and animals.

Chapter 16   Our Earth

Geology

16.1 Shape of Earth

What is the shape of Earth? Is it round like a circle? Is it spherical like a ball? Is it flat like a pancake? Is it square like a block? Does it look like a pumpkin or an eggplant?

When you take a walk around your neighborhood, Earth seems flat. As you go to the park or walk to the store, you don't slide off the ground and you don't lean to the left or right. When you walk on a hill, you go up and then down again. You can walk for many miles and the Earth will seem flat.

However, if you look at the Earth from an airplane or from a boat at sea, you'll discover that the Earth is not flat, but curved. If you looked at Earth from a spaceship, you would discover that Earth is shaped like a ball.

But Earth is not shaped like a perfectly round ball. The middle of the Earth is pushed out a little, and the top and bottom are slightly flattened. The Earth is shaped like a slightly smashed ball!

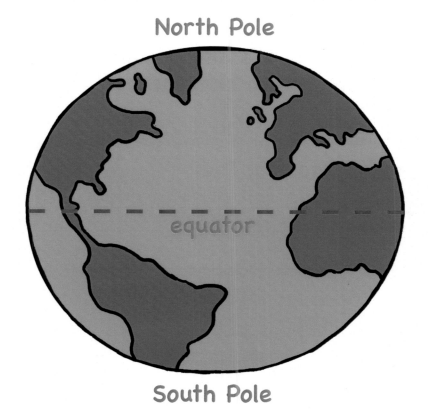

The top part of Earth is called the North Pole, and the bottom part of Earth is called the South Pole. The middle of Earth is called the equator.

16.2 Size of Earth

It's hard to imagine that we walk, run, and live on a planet that's shaped like a ball. If you live at the middle of Earth, near the equator, do you feel like you are standing sideways? If you go to the South Pole do you feel upside down? If you go to the North Pole do you feel right-side up?

In fact, if you go anywhere on the Earth, you don't feel upside down or sideways. You always feel right-side up.

Why does Earth seem flat if it is actually a curved ball? And why do we always feel upright even when we are at the South Pole or on the equator?

Many years ago some people were afraid to travel too far out to sea. They thought the Earth was flat and that they would fall off the edge if they went too far! It made sense to them because the Earth felt flat.

What they didn't know is that the Earth is very large. The Earth is so large that we can't feel the Earth's curve when we walk or run. The Earth is so large that we can't tell if we are on the top, on the bottom, or on the side.

The Earth is about 25,000 miles around. It would take you more than a year to walk around the Earth! That's how big Earth is.

Because Earth is so large, it feels flat when we walk, run, work, or play. But the Earth is really a huge ball.

16.3 Parts of Earth

Earth has different parts. You know that when you walk outside, you step on rocks and dirt. Rocks, minerals, and dirt make up the outer part of Earth. Rocks, minerals, and dirt also make up the ocean floor.

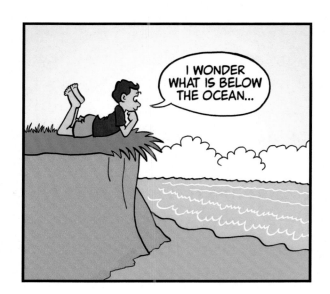

But what is beneath the dirt and the ocean? What is inside the Earth? Is it rocks, minerals, and dirt like the outer part? Or is it chocolate syrup, liquid gold, or melted cheese?

Scientists have no way to actually see what's below the outer surface of Earth, but by studying volcanoes and earthquakes, scientists can come up with some ideas about what lies below the surface.

16.4 Earth's Layers

Earth is a rock planet, which means that it is made mostly of rocks.

If scientists were able to cut the Earth in half, they think they would see at least three different layers. This means that the outer surface of Earth, where you walk, is different from the part just below it.

The crust is the outermost layer of Earth. The crust is made up of rocks, minerals, and soil. The crust is very hard. This hard outer layer supports you when you walk and it supports buildings. The crust also holds the oceans. The crust is relatively thin compared to the layers below it, and it makes up only a small part of Earth.

Below the crust is the mantle. Scientists think the mantle is much thicker than the crust and is made of layers. Scientists believe the outer part of the mantle is hard and rocky. The inner part of the mantle is thought to be softer and hotter. Scientists think this inner part of the mantle might be more like gooey peanut butter and made of melted rock called magma. Scientists think that the magma in the mantle does not always stay in the same place but moves around!

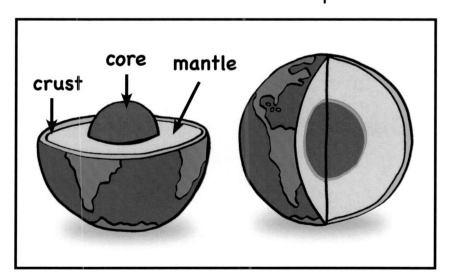

MANTLE

CRUST

CORE

Magma in the Mantle Moves!

Below the mantle and in the very center of Earth is the core. Scientists believe that the outer part of the core is made of molten iron and nickel and the inner part of the core is solid and made mostly of iron. The core is very heavy.

Earth's crust, mantle, and core together are called the geosphere.

16.5 Summary

○ The Earth is made of layers.

○ The main layers of the Earth are the crust, the mantle, and the core.

○ The crust is the hard, rocky, outermost layer of Earth.

○ The mantle is below the crust. Most of the mantle is soft and the outer part closest to the crust is hard.

○ The core is at the center of the Earth. The inner part of the core is likely solid.

Chapter 17 Earth Is Active

Geology

17.1 Introduction

The Earth is an active planet. It is always changing. For example, we can see the surface of the Earth change with the seasons. In the spring and summer some parts of

Earth can be covered with flowers and green grass. In the winter, these same parts might be covered with snow.

You might think that the Earth's crust doesn't change because it is hard and rocky, but it changes every day. There are volcanoes, earthquakes, storms, and flowing rivers that continuously change the Earth's crust.

Geologists study how Earth has changed by looking at rocks, dirt layers, mountains, and other features. Geologists also look at volcanoes, earthquakes, weather, and other things that cause changes on Earth.

17.2 Volcanoes Erupt

Volcanoes can be very exciting. Sometimes when a volcano erupts, a whole mountaintop will come off in a huge explosion!

Volcanoes happen when magma in the mantle pushes up through a weak spot in the crust and comes to the surface of the Earth.

Magma is formed when the rocks and minerals in the mantle melt because the weight of the crust is pushing down on them, creating heat and pressure.

Pressure happens when you squeeze something and it doesn't have anywhere to go. For example, if you keep the lid on your toothpaste and squeeze the bottom of the tube, you will create pressure in the tube. If the top suddenly pops off, the toothpaste will explode! If there is a crack in the wall of the toothpaste tube, the toothpaste will squirt out the side.

This is what happens when a volcano erupts. The magma finds a crack or weak spot in the crust, and because of the pressure in the mantle, the molten rock will spew out!

When the magma comes out onto the surface of the Earth, it is called lava. Lava that has flowed on Earth's surface is responsible for many interesting features. Sometimes lava flows form whole islands. The Hawaiian Islands, for example, are a group of islands made from lava flows.

Volcanoes can form mountains of different shapes and sizes. Some volcanoes form gently sloping mountains over long periods of time. These volcanoes are called shield volcanoes because their long slopping sides make them

resemble the shape of a shield. Shield volcanoes form when very thin layers of lava flow out in all directions. A mountain is formed as the layers build up on top of each other. Shield volcanoes can be several miles long with sides that slope very gradually. The Hawaiian Islands are a series of shield volcanoes.

Volcanoes can also form mountains with steeper sides. Cone volcanoes are cone-shaped because the magma spurts out more quickly and is thicker than the magma that forms shield volcanoes. Also, more rocks and dirt are scattered. The rocks and dirt pile up along the sides of the volcano, making the sides steeper and steeper. Interestingly, cone volcanoes are often found on the edges of shield volcanoes.

Volcanoes can also form dome-shaped mountains. Lava dome mountains are often round in shape and look like a cereal bowl turned upside down! The magma that comes from dome volcanoes is very thick and doesn't flow very far away from the center.

17.3 Earthquakes Shake

Earthquakes can also be very exciting and even scary if you happen to be near one when it happens. Earthquakes occur because sections of the Earth's crust suddenly move. Recall that the hard, rocky crust lies on top of the hard outer part of the mantle, and the hard outer part of the mantle lies on top of the inner part of the mantle that is soft like peanut butter.

Scientists think that the outer part of the mantle is cracked into huge pieces called plates that fit together like a big puzzle. These pieces, or plates, float on the part of the mantle that is soft like peanut butter.

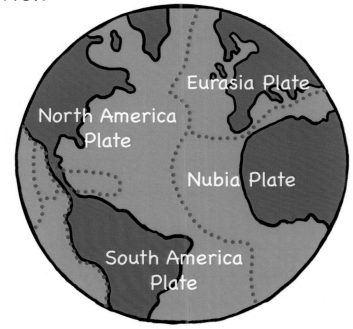

Earthquakes happen suddenly and are usually over within a few minutes. If the earthquake is small, you might feel the floor move, and you might hear a sound like a nearby train. If the earthquake is large, the whole ground moves and buildings and trees may fall. Sometimes after a big earthquake, several smaller earthquakes occur.

When plates move against each other, one section might move upward and the other might move downward. Or the sections might slide past one another. When these things happen, there is movement of the land on either side of the line that is between the two sections.

Let's imagine you have a street in front of your house that separates your yard from your neighbor's yard. Imagine that your house sits on one section of the Earth and your neighbor's house sits on a different section of the Earth. If these two sections move past one another, it's possible that your neighbor from across the street is now your next-door neighbor!

17.4 Summary

○ The Earth's surface is constantly changing.

○ Volcanoes and earthquakes change Earth's crust.

○ Volcanoes happen when magma in the mantle pushes up through a weak spot in the Earth's surface.

○ Earthquakes happen when sections of the Earth's surface move.

Chapter 18 Exploring the Cosmos

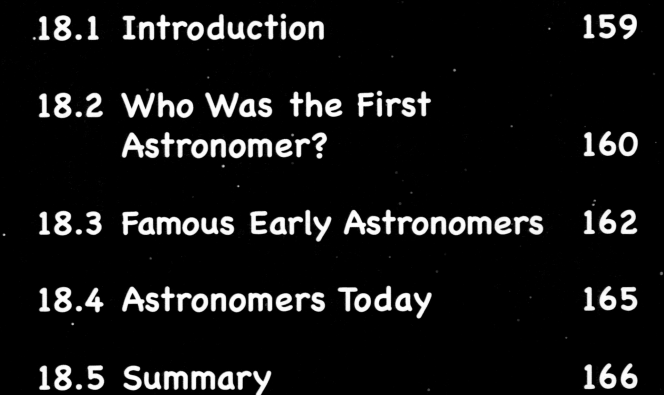

WOO HOO!

Astronomy

18.1 Introduction

So far we have studied four important building blocks of science—chemistry, biology, physics, and geology. We'll now look at the fifth building block of science: astronomy.

Astronomy is the study of the cosmos. The term cosmos refers to the Earth and everything that extends beyond the Earth, including other planets, stars, nebulae, comets, asteroids, and even black holes.

18.2 Who Was the First Astronomer?

Because astronomers can't fly to faraway planets or ride asteroids, astronomers use various tools and techniques

to find out more about the objects in the cosmos. However, before the use of modern tools, people could learn a great deal about the cosmos by studying the night sky.

It's hard to say who was the first astronomer. Many early people studied the planets and stars, and even without modern tools they discovered a great deal about the cosmos.

Early Egyptian, Babylonian, and Mayan people observed the sky in great detail. Noting when the Moon was full or when

the Sun sank lower on the horizon, early observers were able to learn about how the planets and the Moon moved. From their observations they produced calendars and were able to predict eclipses.

One of the questions early astronomers asked was, "Does the Earth move around the Sun, or does the Sun move around the Earth?" In other words, do we live in a "Sun-centered" cosmos or an "Earth-centered" cosmos? To early astronomers it appeared from simple observation that we live in an Earth-centered cosmos. When the Sun rises and sets each day, it has the appearance of moving around the Earth. However, as we will see, sometimes how things move isn't always easy to figure out.

One of the very first astronomers to propose that the Earth moves around the Sun was Aristarchus of Samos. Aristarchus was a Greek astronomer and mathematician who lived from 301-230 BCE. He studied the planets and said that the Earth has two different movements. One movement is that Earth

ARISTARCHUS 301-230 BCE

travels around the Sun, and the other movement is that Earth revolves around its own axis. We now know that he was right! But during his time no one believed him. It would be almost 2000 years before astronomers would look closely at his ideas.

18.3 Famous Early Astronomers

Nicolaus Copernicus was a famous astronomer who also thought that the Earth moved around the Sun. Copernicus was born in 1473 in the ancient Polish city of Torun. During the time Copernicus lived, most scientists believed that the Sun revolved around the Earth. They believed that the Earth was the center of the universe and everything revolved around it.

COPERNICUS 1473–1543 CE

Copernicus did not agree with the scientists of his day. His ideas would eventually change the whole science of astronomy! Unlike Aristarchus, Copernicus was able to use mathematics to show that the Earth moves around the Sun and that the Sun remains fixed in one location. However, Copernicus was not outspoken about his ideas. Because he knew his ideas might upset people, he didn't talk about them. When Copernicus did publish his work, a few people got upset, but most people just ignored his hard work. Another 100 years passed before people took his ideas seriously.

BRAHE 1546–1601 CE

Another famous astronomer also changed the way we see the cosmos. His name was Tycho Brahe. Brahe was born in 1546 in the Danish town of Scania, and he was raised by his uncle. Like Copernicus, Brahe was curious about astronomy. His uncle wanted him to be a lawyer or a politician, but Brahe studied mathematics and slipped away at night to look at the sky. When his uncle died, Brahe was free to pursue his interests in astronomy.

Telescopes that make faraway objects look closer were not yet invented, so Brahe used sighting tubes, which are just hollow tubes with no lenses. Sighting tubes can be used to look at one star at a time. In this way, Brahe discovered that stars do not always appear to be in the same position but are constantly changing. Based on his observations, Brahe decided to rewrite the map of the stars and spent his life working on his ideas.

Galileo Galilei was also a famous early astronomer. He was interested in trying to find out how the planets move. Galileo was born in 1564 in Pisa, Italy. He studied many different subjects, such as mathematics and physics, and he loved to look at the stars. Galileo used his knowledge of math and physics to better understand how the planets and the Moon move.

GALILEO 1564-1642 CE

Like Copernicus, Galileo was an independent thinker, and he didn't believe in an Earth-centered universe. Galileo did experiments because he wanted to show how things moved rather than just coming up with ideas about it. By doing experiments and by using mathematics and physics,

Galileo was able to prove that we live in a Sun-centered solar system that is made up of the Sun and the objects traveling around it. Being able to prove an idea by using experiments, math, and physics was the beginning of astronomy as a science.

18.4 Astronomers Today

Today, many scientists study the stars and planets. Astronomy is a science, and modern astronomers are scientists who use a variety of scientific tools and scientific techniques to learn about the universe.

However, even with new tools, modern astronomers must use the same basic skills that Copernicus, Brahe, and Galileo used.

Today's astronomers must make good observations and must train themselves to see the details, like Copernicus did. Astronomers must also study math and physics like Brahe and Galileo did.

Math and physics are essential for understanding how the stars and planets move in space. Most importantly, astronomers must always be curious and willing to argue to defend their ideas like Copernicus, Brahe, and Galileo did.

18.5 Summary

● Astronomy is the study of space and all the objects found in space.

● Early astronomers were able to discover a great deal about the stars and planets by using observation.

● Nicolaus Copernicus, Tycho Brahe, and Galileo Galilei were three early astronomers who changed the way we understand the universe.

● Modern astronomers still use observation, math, and physics to study space.

Chapter 19 Earth's Home in Space

Astronomy

19.1 Introduction

Now that we know what astronomy is and how to study planets and stars, it's time to explore what Earth looks like from space.

Because Earth is so big compared to our human size, it's hard to imagine what Earth looks like from space. Is Earth the biggest object in space? Is Earth in the center of space? If we took a rocket into space, what would we find?

19.2 Earth Is a Planet

If you launched a rocket and traveled past the clouds into space, you would see the Earth. Earth would look like a blue marble floating in the dark space around it.

Earth is a planet. A planet has special properties. A planet has to be large enough to have its own gravity, which is the force that holds everything to the Earth's surface. A planet also has to move in space around a sun. And finally, a planet is shaped like a ball. Spherical is the word for ball-shaped.

Because Earth is very large, moves around the Sun in space, has gravity, and is spherical, Earth is called a planet.

Earth rotates around an axis, which is an imaginary line that goes through the center of an object. If you were to take a ball and spin it with your fingers, it would rotate around an axis.

The Sun shines on different parts of Earth as Earth rotates. This is how we get our days and nights.

It takes 24 hours for Earth to rotate once around its axis. We don't feel the rotation of the Earth because the Earth's gravity holds down the air and everything else that's on Earth. Everything is moving at the same speed, and the Earth's rotation does not cause wind.

Earth is tilted on its axis. "Tilted" means that Earth is not just straight up and down but is slanted. The tilt of Earth's axis gives us seasons.

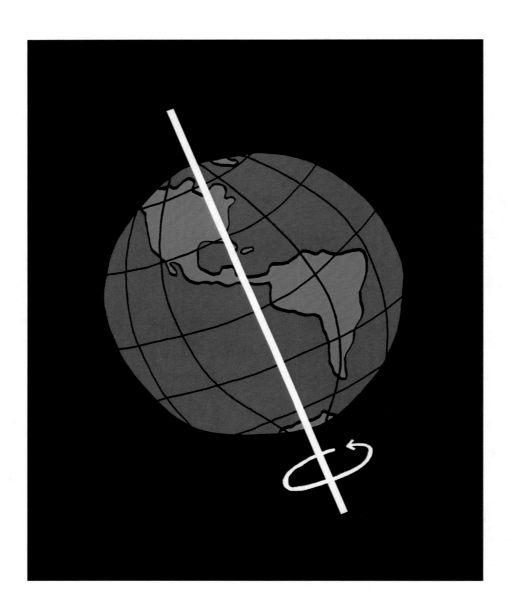

For part of the year, the northern part of Earth is tilted towards the Sun and the southern part of the Earth is tilted away from the Sun. This gives the northern part of the Earth summer and the southern part winter.

Then, during a different part of the year, the southern part of the Earth is tilted towards the Sun and the northern part away from the Sun. When this happens, the southern part of Earth has summer, and the northern part has winter.

19.3 The Moon and Tides

If you take your rocket into space, you might run into the Moon. A moon is an object that travels around a planet. Our moon is smaller than Earth and travels around Earth.

Our moon does have some gravity and is spherical, but because it moves around Earth and not around the Sun as the Earth does, the Moon is not a planet. We will learn more about the Moon in the next chapter.

Did you know that the Moon helps create ocean tides on Earth? It's true! The Moon pulls on Earth with some gravity. This pulling on the Earth causes the water in the oceans to be pulled too.

As the Moon moves around the Earth, it pulls the ocean water with it. The pulling of ocean water by the Moon helps create tides.

19.4 The Sun and Weather

If you wear sandals during a summer day, you can feel the Sun warming your toes. The Sun is a big ball of fire that gives light and heat energy to Earth.

The Earth orbits the Sun. The word orbit means to "revolve around." An orbit is the path one object makes as it travels around another object. If you stand with your hand on a pole and then start walking, you will make a path around the pole. You will orbit the pole. This is what Earth does as it moves around the Sun. We will learn more about the Sun in the next chapter.

Did you know there are storms on the Sun? Did you know that the storms on the Sun can cause storms on Earth? It's true! Sun storms can contribute to Earth storms. Scientists who research weather can study Sun storms to find out how they affect Earth's weather.

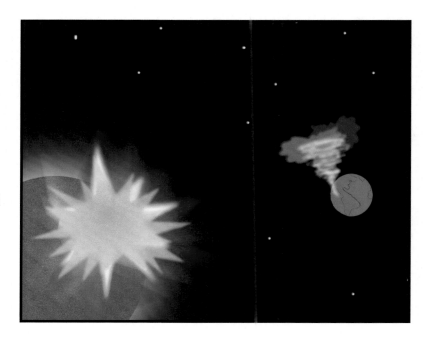

19.5 Eclipses

During the Moon's orbit around Earth, the Moon travels behind the Earth. When the Earth is in between the Moon and the Sun, the Earth can block the Sun's light from reaching the Moon. When the Earth's shadow is cast on the Moon, it is called a lunar eclipse.

At other times, the Moon will be in between the Sun and the Earth. With the Moon in this position, the Moon can block the Sun's light from reaching a portion of the Earth. This is called a solar eclipse.

19.6 Summary

- Earth is a planet.

- One rotation of the Earth on its axis takes 24 hours (one day).

- Earth is tilted on its axis, giving us seasons.

- The Moon and Sun affect Earth's tides and weather.

Chapter 20 Earth's Neighbors: Moon and Sun

Astronomy

20.1 Introduction

In the last chapter you saw that the Earth sits in space. You also saw that the Moon and Sun cause changes in our ocean tides and weather. But what is a moon and what is a sun?

20.2 The Moon

You can see the Moon from Earth. If you look outside your house, you might see a bright, round shape in the sky. This is the Moon.

You can sometimes see the Moon during the day, but most often you see the Moon as the brightest object in the sky at night.

The Moon is spherical like the Earth but much smaller. The Moon has a very different surface from that of Earth.

Although the Moon is made of rocks and minerals like Earth, the Moon cannot support life. It has very little oxygen and no liquid water.

The Moon looks bright in the sky, but the Moon does not generate its own light. Acting like a mirror, the Moon reflects the Sun's light to Earth.

The Moon orbits Earth once a month. If you look at the Moon often, you will see that it appears to change its shape during the month. The different shapes are caused by different views of the Moon as we see it from Earth. The Moon can be round, half-round, or crescent shaped. When the Moon looks round, we say it is a full moon.

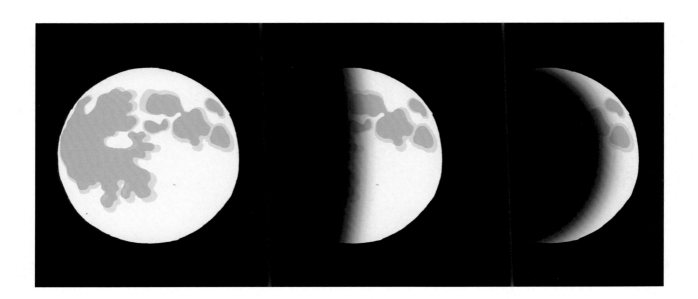

If you look closely at the Moon, you can see light and dark patches. These light and dark patches sometimes look like faces. The "Man in the Moon" is a famous nursery rhyme from Mother Goose.

The light and dark patches on the Moon are actually craters and lava flows.

20.3 The Sun

The other large object you see in the sky is the Sun. You can't look directly at the Sun (that would damage your eyes), but you can see that the Sun is rising in the morning, moving across the sky during the day, and setting at night.

The Sun is not a planet or a moon. The Sun is a star. A star is any object in space that generates its own light and heat energy (see Section 20.4).

The Sun is much larger than Earth. The Sun is so large that a million Earths would fit inside!

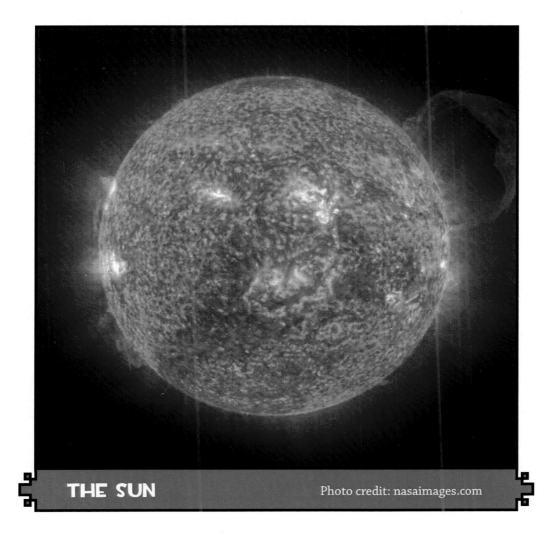

THE SUN Photo credit: nasaimages.com

The Sun is also very hot. The temperature on the Sun's surface is thousands of times hotter than temperatures on Earth. The Sun is so hot that it would melt everything on Earth if Earth were too close to it!

20.4 The Sun's Energy

The Sun generates its own light and heat energy. In Chapter 11 you learned that energy is something that gives something else the ability to do work. The energy made by the Sun comes out as light energy and heat energy. Plants use the Sun's light energy to make food. Making food is work, so the Sun gives plants the ability to do work. We can use the Sun's light energy to make electricity. Our bodies also use the Sun's heat energy to stay warm. Without the Sun there would be no life on Earth.

The Sun is not made of rocks and is not like the Earth and the Moon. Instead, the Sun is made of helium and hydrogen gases. The extremely hot temperatures on the Sun make hydrogen atoms stick together, forming helium atoms.

In Chapter 4 you learned that when atoms stick together they can make molecules. You also saw in Chapter 5 how atoms and molecules rearrange to make new molecules in a chemical reaction. When atoms stick together to make molecules, they use their electrons to attach to each other.

When hydrogen atoms stick together to make helium atoms, they rearrange their protons and neutrons! This is called a nuclear reaction. A nuclear reaction is different from a chemical reaction. In a chemical reaction, the protons and neutrons stay the same and only the electrons stick together.

During a nuclear reaction lots of light and heat energy are released. Because the Sun can generate its own energy, it is called a star.

20.5 Summary

● The Moon orbits the Earth once a month.

● The Moon is made of rocks and minerals, like Earth.

● Our Sun is a star and makes its own energy.

Chapter 21 Earth's Neighbors: Planets

Astronomy

21.1 Introduction

In Chapter 20 we learned about two of Earth's neighbors —the Moon and the Sun. In this chapter we will learn about some of Earth's other neighbors—the planets.

21.2 Planets

If you look into the sky during the day, you can see the bright Sun and possibly a very faded Moon.

But at night, depending on where you live, the sky twinkles up with brilliant specks of light. Many of these specks are stars, like our Sun. But some of the bright lights that dot the night sky are planets.

In Chapter 19 we found out that Earth is a planet. Recall that Earth is a planet because it is large enough to have gravity, is spherical in shape, and orbits the Sun. The Moon is not a planet because it orbits the Earth rather than the Sun.

Earth is one of eight planets that orbit the Sun. The names of these eight planets are: Mercury, Venus, Earth, Mars, Jupiter, Saturn, Uranus, and Neptune.

21.3 Two Types of Planets

All of the planets are different from each other. Mercury has a barren moon-like surface, Venus has toxic gas clouds, and Saturn has brilliant rings. Mars has a reddish color, and Uranus and Neptune look blue or blue-green.

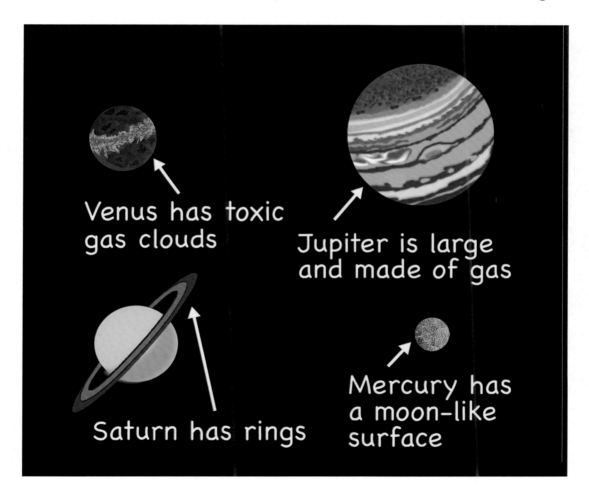

Venus has toxic gas clouds

Jupiter is large and made of gas

Saturn has rings

Mercury has a moon-like surface

Even though all the planets are different from each other, some of the planets have features that are similar. Because of these similarities, scientist are able to separate the planets into two groups. The names of these two groups are terrestrial planets and Jovian planets.

The terrestrial planets are those planets that are most like Earth, and the word terrestrial means "Earth-like." There are four terrestrial planets: Mercury, Venus, Earth, and Mars.

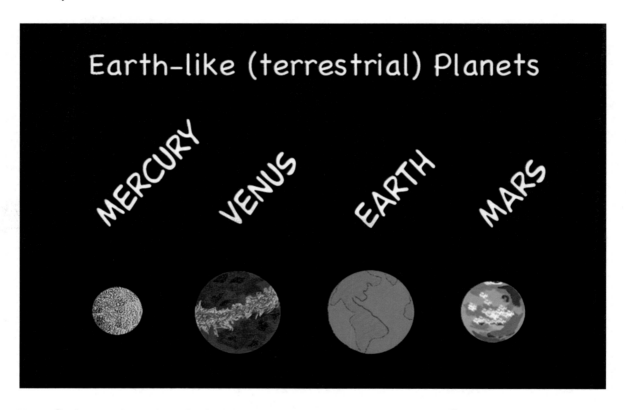

All of the terrestrial planets are made of rock and minerals, like Earth. Also, all of the terrestrial planets have volcanoes, mountains, and craters on their surfaces.

However, of all the terrestrial planets, only Earth has life on it. Only Earth has the water, the oxygen, and the proper atmosphere needed to support life as we know it.

The Jovian planets are those planets that are similar to Jupiter. Jupiter is a very large planet made mostly of hydrogen gas and helium gas. All of the Jovian planets are like Jupiter because they are all very large and made mostly of gas. The Jovian planets are Jupiter, Saturn, Uranus, and Neptune.

21.4 Where's Pluto?

Pluto was once called the 9th planet, but in August 2006 scientists at the International Astronomical Union (IAU) changed their minds. They decided Pluto does not have all the features needed to be classified as a planet. Pluto is now called a dwarf planet or a plutoid.

However, not all scientists agree with the IAU's decision. Scientists argue about conclusions all the time, and arguing is part of science. Someday Pluto may again be considered a planet.

21.5 Summary

- The eight planets are Mercury, Venus, Earth, Mars, Jupiter, Saturn, Uranus, and Neptune.

- The planets are separated into two groups: terrestrial planets and Jovian planets.

- Terrestrial planets are "like Earth." Mercury, Venus, Earth, and Mars are terrestrial planets.

- Jovian planets are "like Jupiter." Jupiter, Saturn, Uranus, and Neptune are Jovian planets.

Chapter 22 Putting It All Together

Conclusion

22.1 Science

In this book you learned that science is a way to study and understand the world around us. You learned that chemistry is the study of atoms and molecules, and physics is the study of force, energy,

and work. You learned that biology is the study of living things, and astronomy and geology are the study of Earth, planets, and stars.

You learned that science gives us information about how things work, what they are made of, how they move, what they do, and whether or not they are alive.

22.2 All the Blocks Fit Together

Chemistry, biology, physics, geology, and astronomy are different parts of science. Like building blocks, they fit together to make science. Because they fit together, you have to learn a little about each building block to understand science as a whole.

You will understand more chemistry when you learn about physics, biology, geology, and astronomy.

You will understand more biology when you learn about chemistry, physics, astronomy, and geology.

You will understand more physics when you learn about astronomy, geology, chemistry, and biology.

You will understand more geology and astronomy when you learn about chemistry, physics, and biology.

The building blocks of science are connected to each other, and together they form the foundation for all science.

22.3 Learning Science

Learning about each of the different building blocks of science will help you understand science as a whole because all of the different science subjects fit together and overlap.

Chemistry explains not only how chemical reactions work in a test tube, but also how chemical reactions work in frogs, batteries, oceans, and the atmosphere. Chemistry also explains how nuclear reactions work in the Sun.

Physics explains not only how work, force, and energy take place in a marble, but also how work, force, and energy take place in butterflies and molecules and during earthquakes and solar storms.

Biology explains how living things live and breathe, and also helps geologists understand how living things change the Earth's surface. Biology helps astronomers guess whether or not life on other planets is possible.

Geology and astronomy help us understand what the Earth is made of and where it is in space. And geology and astronomy also help us explain how changes on Earth, the Moon, or the Sun affect the biology of life.

22.4 Discoveries in Science

If you decide to become a scientist, studying all the building blocks of science will be very important. You will have a much better chance of making new and exciting discoveries if you have knowledge of all the building blocks of science.

By studying each part of science you will be able to make connections you might otherwise miss. For example, by knowing chemistry, you might discover a new molecule important for butterflies. By knowing physics, you might find a new planet in a faraway galaxy. By knowing biology, you might discover a new chemical reaction important for frogs. By knowing geology, you might discover how life can adapt to the extreme environments on a glacier.

22.5 Summary

● Science is made of five building blocks—chemistry, biology, physics, geology, and astronomy.

● All of the building blocks of science fit together.

● Studying all the building blocks of science will help you learn more and will help you make new discoveries if you become a scientist.

More REAL SCIENCE-4-KIDS Books
by Rebecca W. Keller, PhD

Focus Series unit study program — each title has a Student Textbook with accompanying Laboratory Workbook, Teacher's Manual, Study Folder, Quizzes, and Recorded Lectures

Focus On Elementary Chemistry
Focus On Elementary Biology
Focus On Elementary Physics
Focus On Elementary Geology
Focus On Elementary Astronomy

Focus On Middle School Chemistry
Focus On Middle School Biology
Focus On Middle School Physics
Focus On Middle School Geology
Focus On Middle School Astronomy

Focus On High School Chemistry

Building Blocks Series year-long study program — each Student Textbook has accompanying Laboratory Notebook, Teacher's Manual, Lesson Plan, and Quizzes

Exploring the Building Blocks of Science Book K (Coloring Book)
Exploring the Building Blocks of Science Book 1
Exploring the Building Blocks of Science Book 2
Exploring the Building Blocks of Science Book 3
Exploring the Building Blocks of Science Book 4
Exploring the Building Blocks of Science Book 5
Exploring the Building Blocks of Science Book 6
Exploring the Building Blocks of Science Book 7
Exploring the Building Blocks of Science Book 8

Super Simple Science Experiments Series

21 Super Simple Chemistry Experiments
21 Super Simple Biology Experiments
21 Super Simple Physics Experiments
21 Super Simple Geology Experiments
21 Super Simple Astronomy Experiments

Kogs-4-Kids Series interdisciplinary workbooks that connect science to other areas of study

Physics Connects to Language
Biology Connects to Language
Chemistry Connects to Language
Geology Connects to Language
Astronomy Connects to Language

Note: A few titles may still be in production.

Gravitas Publications Inc.
www.realscience4kids.com

GRAVITAS
PUBLICATIONS